ISLAND PRESS
Critical Issues Series

Research Priorities for
CONSERVATION BIOLOGY

Edited by Michael E. Soulé and Kathryn A. Kohm

Published in cooperation with
The Society for Conservation Biology

ISLAND PRESS

Washington, D.C.   *   Covelo, California

© 1989 Society for Conservation Biology

Design, layout, and cover graphic by Kathryn A. Kohm

Special thanks to the National Science Foundation for support of the Workshop and to the School of Natural Resources at The University of Michigan for assistance in organizing the Workshop. Royalties from the sale of this book will go to the Society for Conservation Biology.

Island Press Critical Issues Series: #1

Library of Congress Cataloging-in-Publication Data

Research priorities for conservation biology.

   (Island Press critical issues series)
"Published in cooperation with the Society for Conservation Biology."
   1. Biological diversity conservation. I. Soulé, Michael E. II. Kohm, Kathryn A. III. Society for Conservation Biology. IV. Series.
QH75. R46 1989   333.95   89-15385
ISBN 0-933280-99-8

Printed on recycled, acid-free paper

Manufactured in the United States of America

10 9 8 7 6 5 4 3 2 1

About Island Press

Island Press, a nonprofit organization, publishes, markets, and distributes the most advanced thinking on the conservation of our natural resources — books about soil, land, water, forests, wildlife, and hazardous and toxic wastes. These books are practical tools used by public officials, business and industry leaders, natural resource managers, and concerned citizens working to solve both local and global resource problems.

Founded in 1978, Island Press reorganized in 1984 to meet the increasing demand for substantive books on all resource-related issues. Island Press publishes and distributes under its own imprint and offers these services to other nonprofit organizations.

Funding to support Island Press is provided by The Mary Reynolds Babcock Foundation, The Educational Foundation of America, The Charles Engelhard Foundation, The Ford Foundation, The George Gund Foundation, The William and Flora Hewlett Foundation, The Joyce Foundation, The J.M. Kaplan Fund, The John D. and Catherine T. MacArthur Foundation, The Andrew W. Mellon Foundation, The Joyce Mertz-Gilmore Foundation, The New-Land Foundation, Northwest Area Foundation, The Jessie Smith Noyes Foundation, The J.N. Pew, Jr. Charitable Trust, The Rockefeller Brothers Fund, The Florence and John Schumann Foundation, and The Tides Foundation.

For a catalog of other Island Press books, please write to Island Press, Box 7, Covelo, CA 95428.

About the Society for Conservation Biology

The Society for Conservation Biology is a professional society organized to help develop the scientific and technical means for the protection, maintenance, and restoration of life on this planet — its species, its ecological and evolutionary processes, and its particular and total environment.

In the service of this goal, the Society's objectives include: 1) the promotion of research and the maintenance of the highest standards of quality and ethics in this activity; 2) the publication and dissemination of scientific, technical, and management information; 3) the encouragement of communication and collaboration between conservation biology and other disciplines (including other biological and physical sciences, the behavioral and social sciences, economics, law, and philosophy) that study and advise on conservation and natural resource issues; 4) the education, at all levels, of the public, biologists, and managers, in the principles of conservation biology; 5) the promotion of all of the above through provision of adequate funding; and 6) the recognition of outstanding contributions to the field made by individuals and organizations.

About the Editors

MICHAEL SOULÉ is a professor at The University of Michigan School of Natural Resources. He is founder and current president of the Society for Conservation Biology. KATHRYN KOHM works with The Wilderness Society. She is currently editing a new book on species conservation and the Endangered Species Act.

Table of Contents

Page

ix   Executive Summary

1   Chapter 1
            Introduction

            A. The Discipline of Conservation Biology
            B. Why Basic Research in Conservation Biology is Essential
            C. The Workshop That Produced This Report
            D. The Global Context of Conservation Biology Research
            E. The Primacy of Action
            F. Major Themes and Conclusions
            G. The Organization of This Report
            H. The Audience for This Report: Researchers and Funders

13   Chapter 2
            Ecosystems: Conservation and Restoration

            Priority Areas of Research:
                A. Monitoring Long-Term Trends
                B. Cumulative Impacts
                C. Effects of Climate Change on Diversity
                D. Restoration of Degraded Systems
                E. Land-Use Policy

21   Chapter 3
            The Ecology of Communities and Small Systems

            Priority Areas of Research :
                A. Conservation Strategies for Particular Systems
                B. Ecologically Significant Species for Management
                   and Monitoring

Page

      C. Mutualistic Interactions
      D. Pathogens, Parasites, and Their Hosts

31   **Chapter 4**
      Research in Population Ecology and Viability

      Priority Areas of Research:
        A. Drawing Generalizations from Single-Species Systems
        B. Population Viability Analysis: Population Dynamics
        C. Population Viability Analysis: Population Genetics
        D. The Decision Process
        E. The Molecular Tool Box

47   **Chapter 5**
      Reproduction, Propagation, and Release

      Priority Areas of Research for Animals:
        A. Reproduction and Propagation
        B. Disease
        C. Nutrition
        D. Stress
        E. Behavior
        F. Release Biology

      Priority Areas of Research for Plants:
        A. Plant Reproduction
        B. Field Reproduction and Restoration

55   **Chapter 6**
      Fragmentation

      Priority Areas of Research:
        A. Effects of Fragmentation on Ecosystem Processes
        B. Effects of Fragmentation on Species Interactions
          and Community Structure

C. Effects of Fragmentation on Single Species

65   Chapter 7
        Ethnobiology and Genetic Resources

Priority Areas of Research:
   A. The Effects of Social and Economic Change on
      Biological Diversity
   B. Land-Use Patterns and Biological Diversity
   C. Systems-Level Research and Land Use Patterns
   D. Designing New Production Technologies and
      Land-Use Systems
   E. Relating Economic Return to Conservation Values
   F. On-Site Conservation of Germplasm Resources

73   Chapter 8
        Diversity: Inventory and Systematics

Priority Initiatives:
   A. Identify Critical Areas in Need of Immediate Protection
   B. Produce a Set of Handbooks on Tropical Ecosystems,
      Conservation, and Conservation Biology
Priority Area of Research: Cataloging and Inventory

79   Chapter 9
        **Infrastructure and Training in the
        Developing World**

Priority Initiatives:
   A. Fortifying Private Voluntary Organizations
   B. Short Courses and Graduate Education in the Relevant
      Scientific Fields
   C. Training Programs for Managers: Funding Needs and
      Criteria

Page

D. How Funders of Research Projects Can Facilitate
Training

87   Appendix A
            Land Acquisition and NSF Funding

91   Appendix B
            **The Smithsonian BIOLAT Program
            for Biological Survey**

93   Appendix C
            Workshop Participants

96   **Further Readings in Conservation Biology**

Executive Summary

Conservation biology is a recent response of the scientific community to the wave of global environmental change that is threatening to extinguish a very large fraction of the world's biological diversity. Conservation biologists view their main task as providing the intellectual and technological tools that will anticipate, prevent, minimize, and/or repair ecological damage.

The audience of this report includes foundations, agencies, organizations, and individual researchers who desire guidance on the research and information necessary for effective conservation, both in the short and long term.

Experts from many fields affiliated with conservation biology met for three days in April, 1988, in order to reach some consensus on where the field now stands and on the major, compelling research priorities in conservation biology in the near future. The discussion began with the recognition that research by conservation biologists, including their forebears in the natural resource fields and in overlapping academic disciplines, have already provided managers and planners with important information and planning guidelines. For example, we have learned that merely setting aside habitat will often lead to unsatisfactory results at best, and to irreversible simplification at worst. Among the areas where recent research has lead to changes in the design and management of conservation projects are patch dynamics, edge effects, population size and viability, inbreeding effects, predation, climatic change, pollution effects, ecosystem processes, endemism, migration, and current and potential human interactions.

Because much of this information is already taken for granted in the design and management of protected areas and endangered species programs, we tend to unconsciously discount the value of future research. This we do at our peril. Our understanding of biological diversity, particularly in the tropics, is often surprisingly superficial; we have hardly begun to appreciate the extraordinary complexity of these systems, much less develop long-term protection strategies. Yet the fact that our basic ecological knowledge is far from complete need not, and indeed should not, detract from the immediate utility of the knowledge at hand. This knowledge of natural systems is probably sufficient to establish a system of protected areas that is capable of significantly slowing the erosion of biological diversity.

Nevertheless, the establishment of a comprehensive system of protected areas is no guarantee that such reserves will not suffer massive losses of species as the effects of landscape fragmentation and isolation accumulate over time. As such, we have no choice but to urgently proceed to simultaneously understand the basic mechanisms that fuel and maintain biotic complexity, while quickly proceeding to protect that diversity through the establishment of protected areas. Given the rate of habitat destruction, much of this research must be accomplished within the next few years or decades, at most.

Following are the most pressing and important initiatives and research needs agreed upon at the workshop. The sequence does not imply ranking of importance.

1. **A crash program is needed to carry out extensive surveys and mapping to identify areas that are critical for the protection of natural and genetic resources** because of their high biotic diversity, high levels of endemism, or because of imminent destruction of critical or unusual habitats and/or biotas. These studies should emphasize taxonomic groups that are better known or those that would indicate parallel biogeographic patterns in groups less amenable to censusing. Such an initiative could also generate critical information on the rates of deforestation and other forms of habitat destruction (see Chapter 8).

2. It is particularly important to understand how natural systems "work," especially in the tropics. Therefore **we call for the immediate establishment of a small number of research sites (perhaps four to eight?) in the tropics to develop a coordinated program of comparative research on populations, communities, and ecosystems in relatively undisturbed and secure situations.** We do not call for a large number of such sites, in part because there are too few researchers with the necessary expertise. These focal sites would be especially valuable as sources of long-term, baseline information on global and ecological processes. We further recommend the active participation of local students, professionals, and institutions in this program and in other research projects in the tropics (see Chapter 1).

3. **Studies are also required at all spatial scales to assess the kinds, mechanisms, and magnitudes of impacts on ecological systems.** Here we include the effects of air, water, and marine pollution, the effects of habitat fragmentation, and the effects of biotic mixing through species introductions. These studies should focus on the development and evaluation of alternative means of exploitation and land/water use, with the goal of improving human welfare while minimizing environmental deterioration and the destruction of biological diversity (see Chapters 2, 6, & 7).

4. Studies on the physiology, reproduction, behavior, ecological interactions (including diseases), and viability of individuals, populations, and species have been essential in the protection and management of reserves and other wildlands. **We recommend enhanced support for research that focuses on these fundamental species-level processes, especially with regard to species of critical ecological or economic importance** (see Chapters 3, 4, & 5).

5. Education in conservation biology, wildlands management, and related areas is vital. **Training for both basic scientists and natural resource managers, particularly in tropical, developing**

countries, is sorely needed. Much of this training should occur locally and regionally, benefit local institutions, and strengthen the conservation and management infrastructures in developing nations (see Chapter 9).

For reasons discussed in Chapter 1, the above list omits direct mention of global phenomena that affect both landscape arrangement and habitat quality. These phenomena are of paramount importance for the protection of biodiversity, and are being intensely studied by other groups of experts.

Loggerhead sea turtle hatchling *(Caretta caretta)*

Chapter 1
Introduction

A. The Discipline of Conservation Biology

Never in the history of our species has the biological diversity of the planet been so threatened. Never in the history of the planet has one species so dominated the biosphere and so changed environmental conditions. Conservation biology has emerged within the last decade as a vital component of the social response to the contemporary wave of ecological disassembly. While its practitioners come from the ranks of traditional disciplines, such as ecology, systematics, and wildlife management, to name only a few, the common focuses and concerns are populations, species, communities, and ecosystems that are threatened by habitat alteration, fragmentation, and destruction. Conservation biologists view their main task as providing the intellectual and technological tools that will anticipate, prevent, minimize, and/or repair such ecological damage.

Research is essential if the conservation movement is to proceed effectively. There is always the danger, in other words, of winning the inning—establishing a comprehensive system of reserves through political action—but losing the game for lack of ecological understanding as isolated, island reserves gradually deteriorate through attrition.

The threat of large-scale extinctions is most acute in the tropical regions of the world, which contain most of the earth's known species diversity. Species loss in the tropics is primarily associated with the conversion of natural habitats to human use, a process that results in drastic reductions in habitat and in the fragmentation of habitats that remain (see Chapter 4). The tropics are not the only areas that are rich in species, however. Out of a world total of ca. 260,000 species of plants, southern Africa, like many Mediterranean climatic regions, is extremely rich in endemic species, having nearly 20,000 compared to 28,000 in tropical Africa. China alone has 28,000 species.

This ubiquity of human disturbance and destruction is dramatized by the modern convention of putting quotation marks around such words as "pristine," and "natural." Biologists are painfully aware that there are virtually no unpolluted, unperturbed *sanctum sanctori* left on the planet. Nature, as we observe it today, merely manifests degrees of disturbance. Even remote tropical forests in Zaire and Brazil have shockingly low densities of edible and saleable mammals and birds. Other tropical areas have suspiciously high abundances of plant species that are economically useful to humans. The disappearance of most species of aboriginal large mammals and many large birds coincide with the arrival of humans in Madagascar, New Zealand, and North and South America also may demonstrate how strongly contemporary ecological interactions are influenced by past and present humans. Indeed, there are few sites left, both marine and terrestrial, where one can observe something very close to pre-human nature.

Nevertheless, it is important to understand that protecting biological diversity, as a practical matter, is independent of the pursuit of the Holy Grail of "pristine." Just because a system is not pristine does not mean it is of no value for conservation. The task of conservation is not to preserve some ideal, pristine nature. Rather, its task is to protect diversity.

On the other hand, it is absurd to consider that all human activities are justifiable because humans are "part of nature." More to the point, when humans were sparse and technology primitive, the impact of our species on biological diversity was often negligible, except, perhaps, in the case of large mammals and birds. Employing fire, weapons, and other means, our ancestors caused many extinctions, and they certainly changed the relative abundances of many other species and habitats. Today, however, our species is capable of wholesale elimination of biotas and ecosystems, including those in the seas. In short, the scale of our current ability to destroy is as unprecedented as it is alarming.

The conservation movement, however, cannot accomplish its task merely by arresting the destruction of habitat; there must also be significant improvements in our understanding of:

1. How biological systems work, particularly the interactions that maintain the integrity of biological communities, and the temporal and spatial scales at which they operate;

2

2. How much perturbation they can accommodate;

3. The qualitative and quantitative effects of different kinds, intensities, and geographic scales of disturbance;

4. The patterns in the global distribution of biological diversity;

5. The consequences of fragmentation;

6. The effects of mixing biotas and of introduced species;

7. The reproduction and propagation of selected species;

8. Better ways to integrate human-modified landscapes with more-natural biological communities;

9. Means of restoring degraded ecosystems to minimize human pressures on remaining natural areas and to encourage maximal levels of diversity outside such areas.

B. Why Basic Research in Conservation Biology is Essential

Like other crisis disciplines, conservation biology borrows and adapts ideas and techniques from many fields. Another trait it shares with crisis disciplines is a sense of urgency borne of time limits. Conservation biologists are compelled to study and discover quickly. Natural processes that require large areas and long time periods are disappearing. Soon, in many parts of the world, it will not be possible to study many phenomena because the natural systems where they once occurred will be too fragmented, simplified, and perturbed. Research postponed for lack of funds, particularly in the tropics, means forfeited opportunities to understand and preserve systems that will never exist again. Worse, postponed research means we will not know where much of biotic diversity exists, and what we need to do to protect it before it is lost.

3

C. The Workshop That Produced This Report

The increasing momentum of the conservation biology movement, coupled with its underlying sense of urgency, prompted members of the Society for Conservation Biology and other organizations to call for discussion of research priorities in the field. In response, the Board of Governors of the Society, at its meeting of October 26, 1987, directed President Michael Soulé and Governor Katherine Ralls to pursue funding for a workshop, the results of which would serve as a blueprint for research in conservation biology. Representatives of the National Science Foundation met with several Board Members in April of 1987, and encouraged the Society to submit a proposal for such a meeting. The outcome was two meetings, both of which were funded by the National Science Foundation. First, a planning committee, chaired by Robert M. May, was convened on January 11, 1988.* The workshop that followed was held on Duck Key in Florida, April 16-19, 1988. (A list of participants in the Workshop is provided in Appendix C).

There were two informal resolutions passed at the close of the workshop. Neither of these matters was on the agenda of the Workshop. Nevertheless, both were felt to be of the utmost importance by the participants.

D. The Global Context of Conservation Biology Research

The first resolution dealt with the need to reverse the anthropogenic changes in the atmosphere, stratosphere, and oceans that will have ubiquitous and deleterious biological consequences. Humans are passively engaged in several overlapping, planetary projects that must inevitably alter the conditions for life and greatly diminish biological diversity. We cannot afford to ignore these global problems that are of enormous significance to conservation; for none of them do we have anything approaching full knowledge of their impact on biological diversity.

---

* The other members of the Planning Committee were Drs. Stephen Brush, Robert Hoffman, Thomas Lovejoy, William Perrin, Ghillean Prance, Katherine Ralls, Oliver Ryder, Herman Shugart, Daniel Simberloff, and Michael Soulé.

These macroscale (global) phenomena are of particular urgency for conservation because they affect both landscape arrangement and habitat quality virtually everywhere in the biosphere. Some of these macroscale effects can be classified as habitat destruction. For example, large-scale timber cutting in tropical forests first alters habitat and landscape geometry. In turn, this alters species composition and ecosystem functioning in remaining patches. Such destruction can also affect migratory populations, nearby aquatic and marine communities, and, if widespread, global climate.

Other macroscale phenomena, such as global warming and acid precipitation due to the injection of so-called greenhouse gases and combustion products into the atmosphere, are altering fundamental properties of the environment, such as pH and temperature. Fences cannot protect the biota from these changes. Unless these problems are abated, no amount of research in conservation biology, no campaigns to set aside reserves to maintain biodiversity, can prevent a catastrophic extinction episode.

Multidisciplinary programs (such as the International Geosphere-Biosphere Program sponsored by the International Council of Scientific Unions) are required to examine global processes and how they are likely to effect biological diversity, habitability, and economics. The following problems are among those that need to be addressed:

1. The increasing size of the human population and its accelerating usurpation of the basic nutrient, food, and water supply of all organisms;

2. Rapid anthropogenic climate change caused by increases of CO_2, methane, and other "greenhouse gases" in the atmosphere;

3. Sea-level rise and changes in ocean temperature and circulation, including impacts on coastal zones and near-shore marine ecosystems;

4. Enhanced Ultraviolet light influx as a consequence of ozone depletion in the stratosphere;

5. Widespread air pollution resulting in acid deposition;

5

6. Widespread application of pesticides and fertilizers and their effects on terrestrial and aquatic ecosystems;

7. Widespread ecological change and damage caused by river diversion, impoundment, and siltation;

8. Conflicts between developing and developed countries in economic policy related to biological resources and diversity.

Because these global problems are under intense scrutiny by other bodies of experts, we decided that in-depth consideration at this workshop would be inappropriate. Rather, this report emphasizes the proximal threats to biological diversity—those that are of the scale amenable to research and management interventions at regional and local levels. Many global phenomena, especially pollution of the atmosphere and stratosphere, were considered to be outside the range of this report. We register our deep concern, however, and urge that responsible individuals and institutions take appropriate steps to deal with these global problems before it is too late.

E. The Primacy of Action

The second resolution dealt with the primacy of conservation action. The sense of the meeting was that the Society's first priority must be the protection of biological diversity everywhere. Military metaphors spring to mind in crisis situations. For example, when an enemy invades, the first kind of mobilization is nearly always a martial one. Later, given time and resources, crash research programs (such as the Manhattan Project) may be desirable and even necessary for survival. Just so now; conservation is the most urgent priority, even as we engage in conservation research.

The appropriate balance between conservation and conservation biology (in the broad sense) was not discussed explicitly. Nevertheless, a sense of grave urgency about biological diversity is reflected in the recommendations. Note particularly the recommendations (see Executive Summary) concerning the kinds of innovations and programs that should be implemented immediately for the geographic,

6

systematic, and ecological description of biota, especially in the tropics. The need for such research as a basis for conservation demonstrates the interdependency of action and research and the impossibility of effective conservation in the absence of relevant biological and social knowledge.

Throughout this document, there are references to the importance of long-term research. Such calls for extended studies should not, however, be used to justify delaying conservation actions. The answers that come from long-term research will eventually provide us with a much deeper understanding of the workings of biologically rich and complex systems. This information will assist our descendents in their task of protecting the fragments of biodiversity that remain and restoring as much as possible. In the short term, much can and must be learned if we are to save a significant fraction of our evolutionary tree, especially its tropical branches.

F. Major Themes and Conclusions

Major Threats to Biological Diversity

At some risk of overgeneralization, the threats to biological diversity can be divided into three broad categories: (1) macroscale (climatic and oceanic) stresses, (2) habitat loss and fragmentation, and (3) the effects of introduced species. Issues related to these three categories are noted repeatedly throughout this document, although less emphasis is placed on the first for reasons already stated.

It would be a mistake, however, to assume that all study sites for conservation should be selected because they are either stressed by climatic change, fragmented, or infected with exotic (introduced) species. Perturbed systems cannot provide all the information we need for intelligent management. It is also essential for us to have knowledge of the workings of relatively unperturbed systems.

Priority Initiative: A Worldwide System of Research Centers

The above recognition leads, in part, to a major recommendation of this report. The only way that information necessary for the long-term,

7

intelligent management of tropical ecosystems can be obtained, given current rates of habitat destruction and disturbance, is to identify and establish a series of focal research sites (particularly in the tropics) where long-term, baseline data on ecological processes can be accumulated and where research that is essential for the management of similar areas can be conducted.

This recommendation is not unprecedented. In an earlier authoritative report (Research Priorities in Tropical Biology, NRC, National Academy Press, Washington, D.C., 1980), an expert group of tropical biologists also proposed such an initiative. Among the points made in that report were:

> *We have concluded that the best way to develop a comprehensive understanding of how tropical forest ecosystems operate is to coordinate studies at a few sites that will receive detailed and continuing attention. It is necessary to focus on a few such sites because of the considerable expenditures required and the limited number of trained scientists who could reasonably become involved.*
>
> *[This initiative] will contribute profoundly to the progress of tropical ecology as a science and accelerate fulfillment of human needs.*
>
> *The goals would include:*
>
> * *Obtaining information about the adaptive responses of biota to some of the richest terrestrial environments of the world before the opportunity for study is lost.*
>
> * *Identifying ecosystems (and elements therein) that are found to be most urgently in need of conservation and preservation .*
>
> * *Providing an ecologically sound foundation for assessing and managing the secondary ecosystems that cover most of the tropics.*
>
> *We also assume that biological inventories would be especially intensive at and around each site and that studies related to freshwater habitats, soils, and human populations would be carried out in the vicinity wherever possible. . . .standardization of measurement techniques is vital.*

We would add two points. First, long-term research and the acquisition of land can and should be linked (see Appendix A on land acquisition). Second, it is important to recognize that research is of very great value to the conserved wildland itself for at least the following reasons:

1. The physical presence of researchers and their associates can be a socially powerful deterrent against poaching, squatting, and other forms of encroachment.

2. The research process itself can be a major source of international income for the country and a generator of local income at a scale easily competitive with agriculture (on the low-grade soils and climates commonly associated with conserved wildlands). Long before there are "ecotourist" dollars flowing from a national park, for example, the research there will be a cash generator.

3. The information generated by research can be of very direct value to wildland managers, given that they have adequate training to be able to use it, and given there are effective communications between the researchers and the managers.

Appendix A provides an outline for management-related research topics in these focal research facilities.

G. The Organization of This Report

Chapters 2–5 of this report follow a descending hierarchical sequence; ecosystem processes, community processes, population (small systems) processes, and physiological-behavioral processes. Following these four chapters is a partial synthesis (Chapter 6) on the pervasive effects of habitat fragmentation and the research needed to address these effects. Chapter 7 addresses traditional and contemporary economic uses of biological diversity. Chapter 8 summarizes the roles of systematics and biotic inventories. Chapter 9 focuses on education in conservation biology and management and on the development of scientific infrastructure in developing countries.

Each chapter begins with an introduction to the topic, defining it and outlining the major challenges within it. Next, two chapters have a section "Priority Initiatives" (such as Section F above) that describe and justify actions that should be taken immediately to protect biological diversity or to facilitate research projects with that objective. Most chapters also have a "Priority Areas of Research" section that describes the most important areas of research that will produce the knowledge and technology necessary, ultimately, to protect biological diversity.

H. The Audience for This Report: Researchers and Funders

Researchers

This report includes a distillation of background papers prepared in advance of the Workshop as well as materials written by participants during the three-day meeting, by the editors, and by outside reviewers. It represents an up-to-date assessment of research challenges and opportunities in the broad field of conservation biology. Undoubtedly there are gaps. Overemphases and underemphases reflect, in part, the people who were present at the Workshop and those who were unable to attend, respectively.

It is not the objective of this report to dictate research policy. Not only would that be futile, but we all acknowledge that the anarchy and serendipity inherent in science are often more important than blueprints and written objectives. Nevertheless, it is worthwhile to suggest avenues, issues, and problems that are obviously in need of attention, especially when time is of the essence. This report, in part, is intended to serve as a resource for concerned scientists.

In addition, some problems can be solved only by the joint efforts of many individuals and institutions. This report suggests several such critical research programs. The implementation of these projects cannot wait for spontaneous formation of research groups through the mechanisms of normal science. There must be a strategic process guaranteeing that such initiatives will be implemented in a timely and

10

effective manner. We believe that the economic and biological costs of procrastination will be many times the cost of implementing the recommendations of this report.

Funders

Currently, the National Science Foundation (NSF) and other government agencies fund some of the research in conservation biology. The balance is funded by several non-governmental conservation-oriented foundations and organizations. On several occasions, representatives of both governmental and private organizations have asked representatives of the Society for Conservation Biology for guidance about the major, most pressing research problems in the field. This report constitutes the collective response of 35 or so experts in the field to such requests. Funders should note that "Priority Initiatives" may be especially relevant to the program objectives of private foundations and non-governmental organizations.

Acknowledgments

We wish to thank W. Frank Harris and the National Science Foundation for their support of the workshop that led to this report. Katherine Ralls and Robert M. May assisted in the planning of the project and in the initiation of the NSF proposal. Jean Brennan was responsible for workshop logistics and accounting. The comments of outside reviewers Drs. Thomas Eisner, Terry Erwin, Peter Raven, Reuben Olembo, and Edward O. Wilson on the final draft led to important improvements and clarifications. We also wish to thank the School of Natural Resources at The University of Michigan for their assistance.

deforestation in Brazil

Chapter 2

Ecosystems:
Conservation and Restoration

Individual organisms exchange nutrients and energy with the environment and with each other. On a large scale, such as an entire lake, coral reef, forest, or landscape, the dynamics of these exchanges are referred to as energy flow and nutrient cycling. The discipline that describes these and related processes is called ecosystem ecology.

The importance of ecosystem ecology to conservation biology is based on two premises: (1) ecosystem processes sustain populations, and major ecosystem perturbations change the distribution and abundance of species; and (2) population processes affect ecosystems, and major anthropogenic changes in species diversity alter community composition and ecosystem processes. Many global, anthropogenic phenomena function as perturbations or disturbances. Therefore, the ecosystem research required by conservation biology focuses on the effects of disturbances and falls into three broad categories:

1. What are the functional and spatial characteristics of ecosystems and landscapes needed to sustain viable populations (see Chapter 4)? Of interest here are how the flows of nutrients and water, the dispersal of organisms, and climate are influenced by the geometry of the landscape and by changes within the landscape elements themselves. The development of criteria to identify elements that provide critical habitat for wide-ranging species, or species that are important regulators of material and organism flow between other ecosystems, will be of particular value.

13

2. What are the consequences of species extinction, invasion, and replacement at the ecosystem level? In order to develop and test models that will predict changes in ecosystem processes with changes in species composition, we need a better understanding of how organisms can be classified by functional attributes important at the ecosystem level (i.e., resource response curves, litter quality, effects on microclimate, interactions with higher trophic levels). In other words, conservation biology still lacks a body of empirically based theory or rules about the functional redundancy of species within communities of different types and in different places. It also lacks the basis for generalizing about the amount of simplification in species diversity that different ecosystems can sustain before major changes in ecosystem processes begin to occur.

3. What are the principles that govern the recovery or restoration of highly disturbed ecosystems? Many disturbances that change ecosystems today are different from natural disturbances because of their size, duration, and intensity. Where human activity is particularly intense, the productive capacity of the land is eventually exhausted, and the land is then abandoned. Frequently, these degraded lands are taken over by aggressive grasses or bracken ferns (e.g., *Imperata* in Malaysia, *Paspalum* in Amazonia) and succession is permanently deflected. These depauperate associations (akin to temperate old fields) are a sad contrast to the diverse and stately rain forests that previously occupied the land.

Similar degradative processes occur in semiarid and arid lands as a result of overgrazing and overbrowsing. Many of these degraded and relatively unproductive lands are still farmed or grazed, so the management protocols for their restoration will be the joint products of social science and ecology.

In summary, research is needed on:

- The nature of organism-ecosystem feedbacks;
- Criteria to recognize keystone species (see Chapter 3) with attributes important at the ecosystem level;

14

- The assessment of the impacts of invaders, introduced species, and biotic interchanges;
- The assessment of the impacts of global atmospheric changes;
- The mechanisms of ecosystem degradation, and the means for restoring damaged systems.

Priority Areas of Research

A. Monitoring Long-Term Trends

There is an urgent need for information on the rates at which tropical forests are disappearing and the rates of die-off of temperate forests from air pollution. Among the highest priorities is large-scale, global monitoring of these forests via coarse-resolution remote sensing. These efforts are inexpensive (ca. $100,000/year) and can be repeated at yearly intervals to determine rates of change. They also serve to identify critical areas where finer-scaled efforts should be directed and where protection or preservation activities should be intensified.

In addition to remote sensing, the technology of monitoring may include regional censuses, computerized databases, and analyses of spatial patterns and transitions using geographic information systems. Information emerging from large-scale, long-term ecosystem experiments, such as the Minimum Critical Size of Ecosystems Project in Brazil (sponsored by the World Wildlife Fund U.S.) or the Hubbard Brook project in New Hampshire, can be used to calibrate remotely sensed data from the same region.

In addition, the following questions are worthy of considerable research support:

- How do different species and communities respond to different types of disturbance and other environmental stress? By knowing how different species and communities respond to different stresses, particularly those detectable by remote sensing, we can predict the extent of local species extinction and replacement, and therefore community and ecosystem change.

- How can we better monitor different species in a community on a regular basis over a long time period? There is simply not enough person-power within the ranks of professional scientists to monitor all that needs monitoring. Research is needed to develop the protocols and programs that would allow interested non-scientists (volunteers and conservation groups) to participate more effectively in monitoring the health of our planet. This has worked well with marine mammals; stranding networks documented the mass die-off of dolphins in 1987, and coast-wide, single-day killer whale censuses have been effective. Similarly, it may be possible to modify and improve present censusing schemes (such as bird-banding) so that a wider variety of species is monitored, while important data on life history variables are gathered.

B. Cumulative Impacts

Often, it is not one single disturbance that causes species extinction and ecosystem change, but rather the cumulative effects of many different disturbances. The problem can be simplified, however, if we search for general patterns of change following cumulative impacts. The construction of high dams, for example, often results in predictable impacts, including siltation and an increase in parasitic diseases. The recent outbreak of canine distemper in Baltic seals could be due, in part, to physiological stress caused by the disposal of sewage and toxic chemicals in the ocean. There are other examples of mass mortality in the sea, including the die-off of the sea urchin *Diadema* in 1983-84, and the widespread disappearance of acroporid corals in the Caribbean, which might reflect cumulative impacts, but about which we know very little. As such, research should address the following question:

- Are there repeating patterns of cumulative impact on ecosystems and landscapes that can be predicted from particular types of large-scale development projects?

C. Effects of Climate Change on Diversity

For the reasons given in Section D, Chapter 1, this report does not deal extensively with the causes and effects of anthropogenic global change of the climate and the seas. At this point, we merely mention two very broad categories of research.

- What are the processes of degradation and succession in ecosystems that are likely to accompany global climatic change? For example, several computer simulations suggest widespread forest diebacks followed by gradual invasion by different species as a result of global warming.

- What is the best spatial pattern of reserves to optimize diversity in the face of changing climate? Given the magnitude of habitat loss now occurring throughout much of the tropics, is it possible to design reserves that take into account both existing beta (habitat) diversity and the long-term likelihood that the distributions of many species will shift (often independently) in response to climate changes?

D. Restoration of Degraded Systems

Before discussing research priorities related to restoration ecology, a word of caution is appropriate. Simply put, "fixing" activities will be of little avail if habitat destruction is not halted. Our first priority must

17

be to minimize the amount of the biosphere that ever has to be restored. Secondly, we should make every attempt to prevent further degradation of the productive capacity of land that has already been altered from its "pristine" state.

Enough knowledge is currently available to protect and restore disturbed, though not entirely despoiled, land. For example, erosion control and monitoring can be implemented immediately using agronomic knowledge of both tropical and temperate systems. Management of soil properties such as physical structure, organic matter, cation-exchange capacity, and specific nutrients is already underway and should be expanded.

The U.S. Forest Service has produced simple guidelines for the restoration and management of the cut-over forests of the United States. We envision the development of similar guidelines for field managers that will help them diagnose when the critical processes of a managed landscape are beginning to malfunction. Symptoms may include increased erosion rates, population outbreaks, or invasions of exotic species.

Toward these ends, research aimed at both understanding degraded ecosystems and enhancing their productivity and diversity is a priority. More specifically, the following research questions deserve special attention:

- How do derived grasslands and other degraded ecosystems differ from adjacent natural forests in terms of net primary productivity, biomass, and evapotranspiration (i.e., functional characteristics that affect regional and global climate)?

- Are there thresholds beyond which natural recovery will not occur?

- In the cases of arrested succession (e.g., derived grasslands), what is it that keeps woody plants from establishing (e.g., biotic limitations such as lack of seeds, seed predation, competition for light, nutrients, or water, and allelopathy, or abiotic limitations such as soil strength, fire, and microclimate)?

18

- Based on an understanding of factors limiting woody species establishment, what species attributes would ensure successful invasion of degraded systems, and what is the most cost-effective way of screening the native (and possibly exotic) flora to locate candidate species for (re)introduction?

E. Land-Use Policy

We think that the best way to halt extinction is to convince decision makers, worldwide, that it is in their interest to protect biological diversity. Specifically, we propose a series of interdisciplinary research projects and policy initiatives based largely on analyses of existing data. These should include, but not be limited to:

- An analysis of the cascading ecological destruction associated with roads. Such an analysis would be facilitated by remote sensing. The goal would be a moratorium on the construction of roads in some environments, at least until the environmental costs and benefits have been analyzed. The audience for this type of analysis would be development banks.

- An analysis of measures that could be taken to minimize the amount of land that is cleared for cultivation.

- A compilation of evidence showing that a landscape approach to agricultural intensification within a matrix or network of seminatural habitats can maintain economically stable and sustainable agricultural systems. Such stability has already been demonstrated in intensively managed areas, such as in Europe and Asia. These concepts should be extended to the less intensively managed lands in the tropics and implemented where feasible.

Red cockaded woodpecker *(Picoides borealis)*

Chapter 3

The Ecology of Communities and Small Systems

Assuming that there is a continuum of ecological complexity, the field of community ecology occupies the space between the ecology of single species—population ecology or autecology— and the chemical and physical relations of large sections of the landscape or seascape—ecosystem ecology.

In general, community ecologists are concerned with interactions between species. The traditional focus of community ecology is behavioral-ecological interactions such as mutualism, competition, predation, and disease. The maintenance of species diversity often depends on the persistence of these kinds of interactions. For a given study area, it is therefore necessary to understand the nature of fundamental interspecific interactions. For a given protected area, it is necessary for managers to be cognizant of these interactions and to maintain those that are essential.

Priority Areas of Research

A. Conservation Strategies for Particular Systems

Biological idiosyncrasies uncovered early in the study of a system can suggest appropriate conservation strategies. Three examples can serve to illustrate the importance of such discoveries, as well as management implications of basic research on small systems.

1. Wiregrass, longleaf pine, red-cockaded woodpeckers, and fire are intricately linked in southeastern forests of the United States. The endangered woodpecker nests in old, dying longleaf and loblolly pines; these trees are being replaced throughout the Southeast with young trees for pulp production. Longleaf pine thrives only in association with a ground cover, including wiregrass, which facilitates fires. Fire, in turn, is required for longleaf germination and the elimination of hardwood competitors. However, wiregrass propagates almost exclusively by stolons, so it recolonizes very slowly once removed. In fact, cleared longleaf and wiregrass areas often do not return to longleaf and wiregrass when left unmanaged. Instead, other species dominate the new community. Longleaf pine may also be threatened by loss of crucial interactions with a mycorrhizal fungus that is becoming increasingly scarce. There are data suggesting that the pine is inoculated with the fungus by the fox squirrel, a threatened species in parts of the Southeast. One implication of this work for conservation is that the survival of the red-cockaded woodpecker depends on periodic fire, which in turn depends on a sensitive species of grass.

2. Frogs of the World Wildlife Fund's "Minimum Critical Size of Ecosystems" project in Brazilian rain forest are similarly instructive. Four species breed only in peccary wallows and other small permanent ponds. Thus, conserving these frogs depends on maintaining peccaries or mimicking their wallows.

3. Populations of the large blue butterfly, now extinct in Britain, fell when reduction in grazing depleted its open habitat. The coup de grace was a myxomatosis epidemic that devastated the rabbit population, leading to overgrowth of many remaining open sites. The butterfly requires the open habitat because its caterpillars develop in the nest of an ant unable to survive in overgrowth.

    In each instance, an understanding of the habitat and ecological interactions of the species in question yields necessary information for the conservation of the system. Each system seems idiosyncratic and required intensive study. The pessimism engendered by the apparent slow pace of such detailed studies may be alleviated by the recognition that emphasis on certain types of species will allow more rapid progress than would random sampling of the world's biota.

B. Ecologically Significant Species for Management and Monitoring

Obviously, it is impossible to manage every one of the hundreds or thousands of species that occur in a given area. Intelligent management choices and effective monitoring programs will be based on the results of fundamental research in community ecology and other, related disciplines. The fundamental questions that beg for answers are: (1) what to manage and monitor, and (2) how to manage and monitor?

Recently, attention has focused on three categories of species whose interactions in ecosystems make them especially informative about the quality of communities: keystone species, indicator species, and linking (or mobile link) species.

Keystone Species

Within the last decade or so, conservation biologists have emphasized the critical ecological roles of keystone species. Defined operationally, a keystone species is one that, by its effective disappearance from a system, results (directly or indirectly) in the virtual disappearance of several other species. We feel an urgent need to refine and deepen our understanding of keystone interactions, and of the operation of component systems, especially in the tropics where most of the planet's species exist and are currently at great risk. Current efforts in this regard are grossly inadequate. Listed below are some of the categories of keystone species, as well as examples of the likely consequences of their effective loss from a community.

23

- Top carnivores; dramatic increases in the abundances of prey species and smaller predators, and subsequent overgrazing, overbrowsing, and extirpations of some of the latter's prey species;

- Large herbivores and termites; habitat succession and decrease in habitat diversity;

- Species that maintain particular landscape features; the disappearance of water holes and wallows, beaver ponds, etc.;

- Pollinators and other mutualists; reproductive failure of certain plants;

- Seed dispersers; recruitment failure of certain plants;

- Plants that provide essential resources during times of scarcity; local extirpation of dependent animals, including fruit- and nectar-eating species;

- Parasites and pathogenic microorganisms; population explosions of host species;

- Mutualists with important nutritional and defensive roles for their hosts; increased predation, disease, and dieback of plants.

An understanding of the interactions of a keystone species not only facilitates its direct conservation, but also indirectly assists in the conservation of a significant fraction of an entire community. For example, palms and figs are keystones in a Peruvian tropical forest. Palm nuts escape from all but a group of specialist species that are large and can crush them with powerful jaws or have the ability to gnaw them open . There are not many palm nut consumers, but among them are peccaries and capuchins comprising ca. 30% of the total biomass of fruit-eating species. Figs are heavily consumed by all larger primates, procyonids, marsupials, and many birds. In sum, a group of only 12 plant species (of approximately 2,000) maintains almost all frugivores for three months of the year.

Indicator Species

Indicator species are chosen, in part, because they represent a particular use, ecosystem, or management concern. They are not necessarily keystone species. Indicator species are traditional constructs in wildlife management. The U.S.D.A. Forest Service is required to explain the effects of its potential management practices on indicator species. * Questions concerning which species best serves as an indicator for a particular system remains open for further research.

Mobile Link Species

Mobile link species are species which are important functional components of more than one food chain, plant-animal association, or ecosystem. Logically, mobile link species can be considered to be keystone species, though not all keystone species are mobile links.

The following topics and questions should be viewed as a research scaffolding on which more detailed questions on the roles and uses of keystone, indicator, and mobile link species can be added.

- There is a critical need for researchers to identify the characteristics of these types of species for managers. In particular, it is important to provide protocols for determining keystone, linking, and indicator species for known ecosystems.

- How many keystone species occur in different kinds of communities?

---

* Part of the definition of indicator species used by by the U.S.D.A. Forest Service is as follows: "These species shall be selected because their population changes are believed to indicate the effects of management activities." The definition given here for "ecological" indicator species is broader.

- What are the roles of large predators in different ecosystems? Does the local extirpation of large predators often result in the increase in abundance of smaller predators? For example, the loss of top carnivores seems to have much less effect on herbivore and plant communities in grassland ecosystems (e.g., East Africa) than it does in rain forests (e.g., Manu Park, Peru) or in chaparral.

- It is also important to determine how many keystone species there are in different kinds of ecosystems and how many indicator species need to be monitored to assess the health of a protected area.

- Which species serve as major mobile links between subsets of species within a community and between communities?

- It is important to know whether there are nonobvious linking species.

- Is there some objective way of designating indicator species? Which characteristics of ecosystems are desirable to monitor with indicator species? How many indicator species need to be monitored to assess the health of a reserve?

- Some species (trees) provide "keystone" structural resources. Can we maintain those species that require large logs on the forest floor in harvested systems by leaving logs in place and by leaving some large trees uncut to serve as both sources of nesting sites and future logs during subsequent cutting cycles?

- Finally, we must know how these different classes of species respond to both natural and anthropocentric environmental variability (see discussion of PVA in Chapter 4) and to fragmentation (see Chapter 4).

C. Mutualistic Interactions

An understanding of mutualisms is essential for the long-term perpetuation of diversity. Mutualism ranges from facultative interactions which contribute to fitness but do not significantly regulate the number of either interacting species, through asymmetric interactions, to obligate mutualisms in which neither species can exist without the other. Mutualists are often essential to ecosystem functioning and the dynamics of individual species. For conservation, we need to know:

- Which species in a habitat are obligate mutualists, and which of these are dependent on interactions that are brief or difficult to detect?

- How often are mutualisms substitutable? For example, can an introduced mycorrhiza (root fungus) or pollinator be substituted for native species without change to mutualist populations or ecosystem functioning?

- Are mutualistic interactions more fragile than other interspecific interactions, and hence more likely to be seriously disrupted by fragmentation or variation in climate or other interaction? (see keystone indicator and linking species discussion above).

- Are there patterns in the distribution of mutualism indicating that it is likely to be more frequent in some habitats than others?

D. Pathogens, Parasites, and Their Hosts

The interactions between pathogens/parasites and their hosts are very poorly studied in natural systems. Considerably more long-term field studies and manipulative experiments are needed to increase our understanding of these interactions. It would also be valuable to determine commonalties and differences between the epidemiology of plant and animal host-parasite systems.

- Research is needed on the kinds, structures, and sizes (density thresholds) of populations that are necessary for the maintenance and spread of pathogens and parasites.

- Will certain pathogens and parasites disappear as a result of fragmentation, or will fragmentation simply reduce the probability of disease outbreaks, for example by maintaining host populations below threshold (see Chapter 6)?

- Will reduced genetic variability in isolated host populations increase their susceptibility to current or introduced diseases?

- The above questions alone beg a better understanding of the role of genetic variability in host-parasite relationships.

Although prevention of disease is an important management goal under many environmental circumstances (e.g. zoos), disease outbreaks are probably inevitable in most wild populations; hence

- A better empirical and theoretical understanding of the dynamics of control is imperative.

- The development of molecular techniques to identify infected individuals in the field is needed.

- What are the roles of parasites and pathogens as potential agents of biological control within disturbed (i.e., alien species present) ecosystems?

- Emphasis should be placed on research that will increase knowledge of the extant armory of potential natural agents for biological control, and on preserving habitats of these agents.

- More research is needed on the control of exotic species with disease organisms.

The reciprocal benefits that can be gained by using parasites and pathogens as potential biological control agents of exotic pest species should not be ignored. They preempt the need to produce "man-made" control agents.

Northern white rhinoceros *(Ceratotherium cottoni)*

Chapter 4
Research in Population
Ecology and Viability

Though somewhat arbitrary, this chapter distinguishes those aspects of population biology that emphasize ecological processes per se and those that have direct bearing on our ability to predict whether or not a population will be able to persist. The former area (ecology) considers such topics as rarity, distribution, dispersal, and the impacts of introduced species. The latter area (viability) considers demography, genetics, and environmental variability from the perspective of the biologist desiring to assess probability of persistence.

Priority Areas of Research

A. Drawing Generalizations from Single-Species Systems

Understanding the dynamics of populations is a cornerstone of conservation biology. If we do not conserve populations, we will not save biodiversity. Nonetheless, intensive study of the dynamics of populations of selected species and small groups of species is woefully underfunded. Such research serves two broad goals: (1) it allows development of sound conservation strategies for particular systems, and (2) it promises generalizations of potentially wide applicability. Such research must usually be long-term if it is to generate data that will enable detection of a signal (such as declining population size) over the noise (or random fluctuations) inherent in most systems. One might

31

expect that conservation recommendations based on intensive short-term studies may have to be reconsidered in the light of subsequent information; rare events such as a drought or an epidemic may be key to the long-term population dynamics of a species.

Furthermore, although much research underlying conventional wisdom in ecology and population biology focuses on a single species or small groups of species, there has too often been a bias toward studies of abundant species rather than those that have moderate population sizes or are rare. The systems or species chosen for study are too often chosen because of ease of manipulation, aesthetic taste, or economic importance (e.g., crops and pests have received heavy emphasis). A shift of focus could dramatically increase information crucial to conservation. For example, research on threatened species (in which the research itself does not unduly increase the threat), though logistically difficult, may be more directly applicable to conservation efforts. Similarly, research on keystone species in threatened systems will be doubly rewarding. In addition, certain entire taxa have received disproportionate attention (often as a matter of taste), whereas others (such as fungi and nematodes) which may be of great ecological significance are particularly poorly understood. Appropriate distribution of effort would at least partially redress this imbalance.

Most conservation problems reduce to the temporal and spatial processes: the population dynamics of a species and the dynamics of its geographical range. Extinction consists of either the population size or the range going to zero. Thus:

- Research leading to generalizations about population and range dynamics would greatly aid conservation. To the extent that the research organisms were themselves of conservation interest, important management information would arise even before generalizations were clear.

Rare vs. Common Species

Threatened species are often rare; yet, as noted above, most research is on common species. Therefore, research is needed on the following questions:

32

- Do rare species share certain traits? If so, are these traits the causes of their rarity, or of evolved responses to it? Research that compares rare species and related, common species would be particularly enlightening. Some species appear always to have been rare and, as long as their range is not so limited that a single catastrophe could eliminate them or significantly reduce their genetic viability, there may not be grounds for viewing them as threatened.

- Do traits shared among rare species suggest particular conservation strategies? For example, rare prairie grasses appear to do particularly well in experimental competition with sympatric common species; this trait suggests that establishing small monocultures of such grasses would be an inefficient conservation strategy.

- Of particular interest would be studies on recently rare species which may be relatively inviable at low population densities.

Geographic Range Limits

A related important area of research concerns the determinants of geographic range limits—why do densities of a species become low to the point of vanishing toward a range boundary? Such research is particularly timely in view of the prospect of global warming. Global warming overlain by more complicated local climatic variation is likely to change the range limits of many species. In the Northern Hemisphere, small, isolated populations at the southern terminus of the range of a species are worthy of attention.

- To what extent are range limits set by temperature or other physical factors, and how rapidly can physical tolerances evolve? What would be the consequences to a dry tropical forest, for example, of a two-degree increase in temperature?

Such whole organism physiological research can provide essential

33

information for conservation. For example, the northern range limit of Gambel's oak was studied by transplanting seedlings to different microhabitats. The investigators determined that the upper elevation limit is set by temperature stress, while the lower elevation limit us set by moisture stress. These two limits converge northward until suitable microsites disappear. Recent work on checkerspot butterflies, plants that grow on mine tailings, and industrial melanism highlight the extreme sensitivity of the population dynamics to subtleties of climate and substrate variation.

However, because there are so few studies, it is difficult to forecast conservation consequences of physical factor change and the likelihood of sufficient evolutionary response. Though it appears that the prospects for transplanting highly local species, apparently adapted to narrowly defined physical conditions, are poor, there is little literature on the nature and genetic basis of the adaptations.

Long-Term Changes in Species' Ranges

A related sort of single-species research with potential conservation uses focuses on long-term changes in species' ranges. One window into the future is the past. Is an observed contraction of 10 km in one century outside of the "normal" range of variations observed over the last few thousand years? Or more generally,

- In what ways are long-term range changes associated with climate, and to what extent are they limited by dispersal? What are the portents of the impending global climatic changes?

The contents of packrat *(Neotoma)* nests for the late Pleistocene suggest that the ranges and abundances of many species in southwestern plain communities have undergone major fluctuations and that these have been largely independent. The more general question raised by this finding is:

- To what extent are changes in the range of different species linked to functional interactions of the species?

The Effects of Introduced Species

Finally, effects of introduced species (and potential effects of genetically engineered organisms) are of increasing conservation concern. There are too few detailed studies of particular introductions, or comparisons between successful invaders and related unsuccessful ones, for cogent generalizations about the determinants and effects of success

- Intensive studies of particular introductions would provide predictions and management guidance in particular cases and would eventually lead to useful generalizations.

That prediction is possible, given intensive analysis of a small system, is demonstrated by a study that anticipated the trajectory of a planarian introduction into the U.K., including the identity of the native species most likely to be affected, and the particular habitats where the invader would most likely be important.

Dispersal

Research should be directed at the role of dispersal in species extinction and invasion. For example,

- Much more information is needed about the effects of topographic barriers and corridors on dispersal and about the minimum or optimum topological requirements for these landscape features with respect to different species.

- In shallow water marine environments, effective conservation of particular populations will depend on a knowledge of larval life history and dispersal.

Figure 1

The general expected relationship between population size and average persistence time. Solid and broken lines indicate natural and captive populations, respectively.

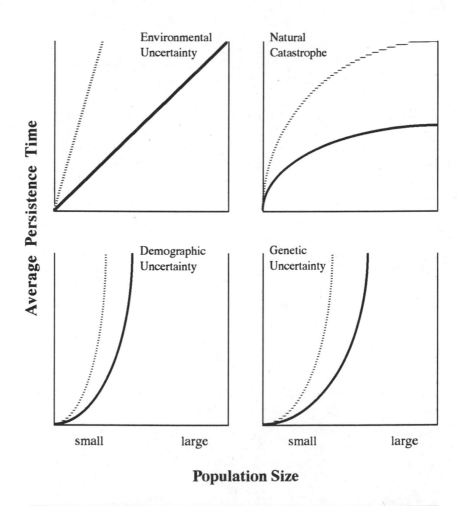

B. Population Viability Analysis: Population Dynamics

Let us assume that an area of suitable habitat will be set aside for a given species. To determine the size of this area, we first need to know the necessary population size (or minimum viable population size, MVP) to avoid extinction of this species due to the various chance factors affecting its population dynamics. More precisely, the MVP is the population size which provides a given probability of persistence of the population for a given amount of time (e.g., a 95% expectation of persistence without loss of fitness for several centuries).

Determination of the MVP for a given situation is a complicated problem and must include the various chance factors that may cause extinction. These include factors that are extrinsic to endangered species, such as environmental uncertainty (variability) and natural catastrophes, as well as intrinsic factors, such as demographic and genetic uncertainty.

For each of these factors, we can graphically relate the average persistence time of a population to its size (Figure 1). The solid and broken lines indicate this relationship in general terms for natural and captive (or highly managed) populations, respectively. First note that in natural populations of large size, environmental uncertainty and natural catastrophes appear to be the most critical factors. In small natural populations, all four factors may be important. On the other hand, in captive populations of small size, probably the most important factors are the intrinsic ones, demographic and genetic uncertainty.

Because these factors may interact, the approach called population viability analysis (PVA) was introduced to consider the complex inter-action of factors influencing extinction. For example, an environmental perturbation may reduce the population size and induce both detrimental demographic and genetic changes which in turn can reduce the number of patches occupied by the population. These connections are often hard to evaluate explicitly or to predict. To date, they have not been acceptably integrated into a model of population persistence.

Theoretical work to date indicates that the probability of a population's persistence is most sensitive to the level of environmentally induced variation in the population's vital rates (e.g., birth and death rates). Indeed, attempts to apply models of population persistence for a range of mammalian species (based on various indirect estimates of population densities, growth rates, and levels of environmental variability) yield estimates of population sizes necessary to achieve a high probability of persistence (95%) over the medium to long-term (100 to 1,000 years) that range from thousands to hundreds of thousands of individuals—numbers that are many times higher than we might expect.

Should these estimates prove true, the prospects for conserving many mammalian species, in situ, at current levels of effort (e.g., current number, size and distribution of parks, game reserves, etc.) are extremely dismal.* However, determining whether these projections are true hinges critically on the answers to a number of basic research problems. Foremost among these is development of truly integrated population persistence models, either analytic or simulation, that are capable of simultaneously tracking the impacts of various chance factors (demographic, environmental, and genetic uncertainty, and natural catastrophes).

Even more important is the development of such a model that is geographically structured (i.e., a metapopulation model). It may, in fact, prove true that a certain magnitude of geographical structure is essential for a reasonable prospect of survival. But this question cannot be adequately addressed with current population dynamics models.

Any model, however, is only as good as its assumptions. Therefore, a much more extensive examination of the relationship between the variability in environmental parameters and the variability of birth and death rates in natural populations is no less important than the development of more sophisticated and realistic models.

Not only is it important to understand the basic connection between environmental and demographic changes; it is also important to understand the geographical structure of environmental variability. How does the degree of variability in one environmental input relate to others, and over what range is such variability consistent (i.e., can we

---

* A necessary caveat is that such huge numbers are probably unnecessary if the population is subdivided in sites that are unlikely to be affected adversely by the same perturbations.

map isoclines of variance ranges for key environmental parameters)? In summary, an increase in the sophistication of PVA will depend on progress in four major research directions relating to the study of population dynamics and persistence, namely:

- Development of an integrated model of population persistence that simultaneously tracks the effects of variation in demographic parameters and in the frequency and amplitude of environmental, catastrophic, and genetic variation;

- Expansion of such a model to incorporate geographic structure;

- Examination of the relationship between levels of environmental variability and their impact on the variability of key demographic parameters for a variety of representative species;

- Examination of the geographic patterns of environmental variability itself.

The approach to these major research priorities should emphasize and be based upon long-term, detailed studies of single-species systems. Certain large-scale field and laboratory experiments could also be employed. Such studies should emphasize the integration of relevant disciplines and the simultaneous examination of both demographic and genetic characteristics and processes (the latter are discussed below).

C. Population Viability Analysis: Population Genetics

An important concern in endangered species conservation is the lowered fitness resulting from inbreeding and the loss of genetic variation from small population size. Research is needed in the following areas:

Inbreeding Depression

Inbreeding and inbreeding depression may not be completely correlated. For example, for a given increase in inbreeding, fast inbreeding may result in a larger decrease in fitness than slow inbreeding. In theory, the probability of fixation of detrimental alleles in a finite population may be used to predict this effect. In fact, if we know the composition of the genetic load (see below), the tolerable rate of increased inbreeding may be determined.

- What is the basis for the one- to three-percent permissible increase in inbreeding per generation suggested by Franklin and Soulé? When, if ever, should a program to purge lethal alleles be recommended?

Inbreeding depression (in *Drosophila*) is caused (in nearly equal amounts) by lethal alleles that are nearly recessive and detrimental alleles of much smaller effect that are not so recessive. The lethal load may be removed more easily by inbreeding while some detrimental load may still exist at fairly high inbreeding levels.

- What is the constitution of inbreeding depression in other species?

It is important to know the effect of inbreeding on total fitness. For example, complete homozygosity for an autosome in *Drosophila melanogaster* results in viability that is only 50-60% that of a heterozygote. On the other hand, the net fitness of a homozygote which includes other components of fitness besides viability is about 15-20% of a heterozygote. In addition, inbreeding depression may be density- or environment-dependent, either factor possibly being different in captive or optimal conditions. Research questions include:

- What is the relative importance of other fitness components on inbreeding depression?

Figure 2

The frequency of selfed- and random-mated progeny having different heterozygosities, and the fitness for different heterozygote classes when a sample is composed of 25% selfed progeny and 75% random-mated progeny, and the fitness of selfed progeny is half that of random mated progeny.

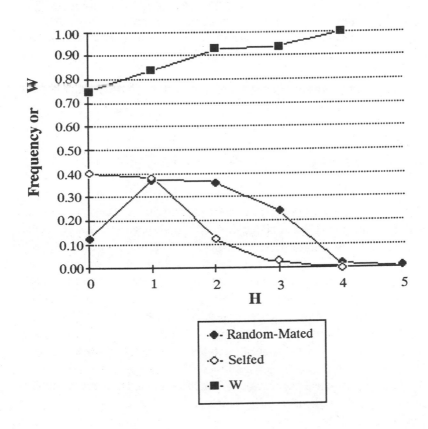

- Under what environmental conditions should inbreeding depression be measured?

Population Bottlenecks

It is generally agreed that population bottlenecks should be avoided as much as possible in endangered species. In nearly all documented cases, bottlenecks result in loss of genetic variation. (Apparent exceptions exist, but even these give evidence for lowering of mean phenotypic values and fitness.) More experimental (and theoretical) work needs to be done to determine the generality of these findings, particularly for fitness components. Reduction in mean fitness may be the most important factor to monitor in a genetic conservation project.

- What are the effects of bottlenecks on mean fitness and genetic variation?

Relationship of Heterozygosity and Fitness

A number of biologists suggest that heterozygosity as measured by electrophoresis is a good measure of individual fitness or population variation in fitness (potential for inbreeding depression). Yet the actual cause of an association between fitness and individual heterozygosity may not be any intrinsic properties of the loci (or associated linked region) examined. Such an association may stem, for example, from the composite nature of the sample, which may include some inbred individuals. In Figure 2, selfed progeny which consist of 25% of the sample, have half the fitness of random-mated progeny. There needs to be an objective evaluation of such associations, as well as an effort to understand the causal mechanisms.

One can construct scenarios in which there are all combinations of heterozygosity and inbreeding depression (Table 1).

- Is it more likely that there is a positive association of heterozygosity and inbreeding depression? When does this occur?

Table 1

Situations that may result in various associations of heterozygosity and inbreeding depression

Heterozygosity	Inbreeding Depression	Situation
High	High	Large population at equilibrium
Low	Low	Population shortly after severe bottleneck
Low	High	Population some time after severe bottleneck
High	Low	Mixture of two populations after severe bottleneck, some level of inbreeding in every population

43

In summary, to better understand the application of population genetics to conservation biology,

- We need detailed examination of genetic phenomena such as inbreeding depression and population bottlenecks.

- We need to examine the relationships between measures of genetic variation and fitness (e.g., heterozygosity and inbreeding depression, genetic distance, and outcrossing depression).

D. The Decision Process

Alternative measures for enhancing viability and for the mitigation of disturbance (development) are often considered by experts who are planning the management or recovery of endangered species. With this in mind,

- Further research is needed on the utility of decision analysis as a means of clarifying the logic of these deliberations.

E. The Molecular Tool Box

We wish to emphasize that molecular biology has a great deal to contribute to conservation biology. For population questions, for example, modern molecular genetics provides an opportunity to obtain more accurate and direct answers to familiar questions. A primary problem is identity. DNA can easily identify members of a family, deme, population, and species, providing better information about diversity at each level of organization. DNA sequences can also illuminate interactions between species, for example, the relative specificity of a pathogen for its hosts, recent and historic. Genetic exchange between close and distantly related organisms can also be monitored. Molecular genetic techniques can track gene flow and dispersal within and between ecosystems. Finally, ecological perturbations can be studied as they affect gene frequencies through time and space.

Thus analysis of information at the level of DNA will enable conservation biologists to better describe the genetic patterns and mechanisms underlying species interactions.

California condors *(Gymnogyps californianus)*

Chapter 5
Reproduction, Propagation, and Release

Captive populations are valuable resources for basic research in genetics, behavior, reproductive physiology, and conservation education. In the animal kingdom, "charismatic megavertebrates," such as rhinoceroses, tigers, and condors, have been the major focus of this research. These species are particularly vulnerable to extinction due to their need for large areas of habitat and their susceptibility to hunting and other human activities. In many situations, efforts concentrated on their survival facilitate the conservation of less dramatic and attractive species.

In the plant kingdom, species do not usually require as much habitat space as large animals. Nevertheless, research relating to the reproduction, propagation, and release of plant species is vital to efforts to conserve biological diversity. In short, as the worldwide destruction of habitats continues, more and more plant species will have to be propagated in botanic gardens or stored in seed banks.

Already, captive populations preserve some species that are extinct in the wild, such as Pére David's deer and California condors. The number of such species can be expected to reach the thousands within the next century. Although some species may have to be maintained indefinitely, due to the destruction of an entire natural habitat, the ultimate goal of many captive-breeding programs is to provide plants and animals for reintroduction into the wild.

Methods for maintaining many species in captivity are still unsatisfactory due to the lack of basic knowledge about various aspects of the biology of these species. Similarly, there has been almost no systematic study of reintroduction techniques.

Priority Areas of Research for Animals

A. Reproduction and Propagation

Successful reproduction and propagation of species in captivity depends on a sound understanding of reproductive biology obtained through behavioral and physiological research. Research priorities tend to fall within at least one of three categories:

1. Observational studies establish basic behavioral patterns associated with sexual activity and the rearing of reproductively competent young.

2. Artificial insemination, embryo transfer, in vitro fertilization, and the stress-free monitoring of reproductive activity are techniques which are immediately useful for basic research and enhanced propagation.

3. The acquisition, transfer, and cryopreservation of germplasm will enormously reduce the capital requirements for breeding facilities while providing insurance against loss of species as a result of natural catastrophes. More importantly, frozen sperm and embryos could be used interactively with living populations to periodically infuse managed groups with genetic material from previous generations and from a much larger number of individuals than can be maintained in captivity or even in parks and reserves.

B. Disease

Because disease pathogens compromise propagation, they warrant serious attention. Research into the impact or elimination of diseases is mandatory at three levels: (1) ensuring survival and reproduction of individuals caught in the wild and maintained in captivity, (2) avoiding infectious agent transmission to other captive stocks, domestic species, or humans, and (3) eliminating the possibility of creating native population epizootics after the release of individuals bred in captivity. More specifically,

- There is need for a database of normal physiological reference values for both disease-free and disease-affected species.

- Basic research is needed on the detection, evaluation, and etiology of pathogenic states potentially catastrophic to captive and free-ranging populations.

C. Nutrition

The food resources of vertebrate species in natural habitats vary through time. This diversity and opportunity for choice cannot be duplicated in captivity. Standard diets have been formulated for many species with similar food habitats (carnivores, seed eaters, nectar feeders, grazers, fruit eaters, and some browsers) and are routinely used in zoos and animal research facilities. However, research is needed on a continuing basis to define and refine the nutritional requirements for a wide variety of species. At present, for example, diets for insectivores and folivores are inadequate.

D. Stress

All organisms have mechanisms for responding to novelty, threats, and disease (stressors). For vertebrates, prolonged activation of these

mechanisms can result in behavior anomalies, tissue damage, reproductive failure, and death. Many species appear to suffer from chronic
stress in captivity.

- Research is needed on variations in these stress mechanisms in
  model species and on ways to prevent deleterious effects.

- Basic research is needed on the impact of stress on the well-
  being and fitness of wildlife species, including the effects on
  behavior, endocrinology, nutrition, and immunology.

E. Behavior

Progress in propagation will be accelerated through comparative
studies of captive and free-ranging wildlife populations. The basic
biological characteristics of a species can be used to develop master
plans for species survival by improved natural propagation and, if
necessary, by artificial breeding. Based on present data, there is little
doubt that species-specific conditions and traits will necessitate sub-
stantial species-specific research.

F. Release Biology

The reintroduction of animals into their historical habitat is becoming an increasingly common tool in conservation. The existence of
more than one population greatly increases a species' chances for long-
term survival. Some reintroduction efforts have used animals captured
in the wild, such as big-horn sheep or California sea otters. However,
reintroductions of captive-bred animals are increasing. Recent examples include the Arabian oryx, the golden lion tamarin, the Bali
mynah, and the Guam rail.

There are few guidelines for the successful reintroduction of species into relatively natural settings. Appropriate recommendations for different species must be developed by a series of well-documented releases of model species.

- Genetic criteria are needed for selecting individuals to be released.

- Guidelines are required for rearing techniques (e.g., hand-rearing vs. puppet rearing vs. parent rearing), and animal selection (e.g., for age, sex, social group). In addition, specific and general guidelines are needed for release sites and techniques, post-release protection and dietary supplementation, and for the amount and type of pre-release training to overcome behavioral deficits.

- The effectiveness of release programs should be evaluated by long-term, post-release studies focusing on the reproductive success and demography of the new or enhanced population.

Priority Areas of Research for Plants

Among the central problems in plant reproduction and propagation is the identity of pollinating and dispersing agents, particularly keystone species (Chapter 3). Successful long-term maintenance, both in nature and captivity, depends on the acquisition of knowledge about these relations.

A major problem in the captive reproduction of plants is the viability of seeds following their collection. Other problems are: (1) the number of seeds that should be collected, (2) the storage of seeds, and (3) the requirements for germination. During propagation, it is necessary to establish the proper seedling requirements for light, nutrition, water, the control of pathogens, and the presence of mycorrhizae and *Rhizobium*.

A. Plant Reproduction

Requirements for successful reproduction are highly species-specific. Therefore, emphasis should be placed on those species with economic value and/or restoration potential, as well as rare species in general. For captive reproduction of plants, research on several factors is needed:

- What are the natural pollinating and seed dispersing agents?

- How many parents should be used in a cross-pollination program, and from what geographic range should they come?

- What number of seeds should be collected; should they be collected to ensure capture of a high number of viable seeds prior to natural dispersal?

- How viable are the seeds, and what storage conditions are optimal for long-term viability?

- What are the requirements for successful seed production (pollination), seed storage, seed dispersal, seed germination, and seedling establishment for different species?

- What are the seedlings' requirements (e.g., light, nutrition, water, control of pathogens, presence of mycorrhizae and *Rhizobium*)?

B. Field Reproduction and Restoration

- During restoration, what are the relative merits of sowing seed directly (by hand or by plane) vs. transplanting nursery-grown seedlings?

- How can seed predation and disease be minimized during regeneration attempts?

- How are pollination and dispersal (hence gene flow/inbreeding) affected by plant population size (rare vs. abundant) and spatial pattern (high vs. low density)?

- How are seed production and the genetic structure of plant populations affected by changes in the population size of pollinators and seed dispersal agents?

Aerial view of isolated forest reserves, Brazil

Chapter 6
Fragmentation

Fragmentation of habitats is one of the major threats to biological diversity (Chapter 1). It exerts its effects through both habitat loss (as in the case of the rapid destruction of tropical moist forests) and habitat insularization (the isolation of habitat patches as a result of development, roads, timber harvests, etc.). One of the central challenges of conservation biology is to measure the nature and rates of anthropogenic ecosystem fragmentation and determine the implications for the loss of diversity.

The implications of current worldwide loss and fragmentation of many habitat types are severe. Even under the most optimistic predictions, we only can hope to preserve a small percentage of the world's species in parks and reserves; the remainder will survive, if they survive at all, outside protected areas. The problem is most serious (by one to three orders of magnitude) in the tropics. Most of the world's species are tropical, and the scale at which tropical ecosystems are being destroyed is unprecedented.

We already know in a qualitative way that habitat fragments lose species; what is not known is the generality of the rules that determine why some species are lost more quickly than others or the extent to which these rules can be predicted from the structure of the intact community and the species interactions. To understand these rules, a comparative program of research is needed to improve the predictability of species loss from fragments. We hasten to add, however, that the generality and utility of the results of such research depends on related studies in control sites that are relatively unperturbed. Such studies are necessary to establish the baseline conditions and ecological relations of biotic diversity in relatively intact communities.

Baseline studies in control sites are also needed to understand and predict the effects of fragmentation on ecosystem processes. Clearly, fragmentation will affect the flow of matter at ecosystem boundaries, but this poses formidable problems with regard to sampling; variability of nutrient, energy, and water flows are greatest at the interface between two ecosystems. In most cases, fragmentation of the landscape will enhance nutrient and water transfers between ecosystems. In addition, fragmentation can alter the spatial patterns of seed dispersal, thereby enhancing the invasion of some ecosystems by some species and decreasing the dominance of others. The ecosystem fragments will then be altered to the extent that the invading species can influence nutrient cycles and carbon storage. Natural landscapes are often highly fragmented, but scarcely anything is known about how and why boundaries form between their component ecosystems. Knowledge of these processes could be incorporated into resource development programs to minimize some of the more drastic effects of fragmentation on material and organismal flows across the landscape.

Therefore, we reiterate the call in Chapter 1 for the establishment of a system of research sites and stations, mostly in the tropics. Comparative baseline research must be initiated immediately at these facilities.

Priority Areas of Research

A. Effects of Fragmentation on Ecosystem Processes

The line between ecosystem and community processes is arbitrary. For the purposes of this report we include the following as ecosystem processes that are likely to be influenced by fragmentation: energy flow within and between habitat patches, nutrient cycling, hydrology, patch dynamics (succession and distribution of habitat patches in space and time) within fragments, canopy development, and the movement of materials across landscape boundaries. Research questions related to these processes include:

- What determines the establishment of natural boundaries between ecosystems, and how can this knowledge be used to minimize detrimental effects of anthropogenic fragmentation?

- How are material transfers from and into ecosystem fragments related to the size and geometry of the fragments and the interspersion of fragments across the landscape?

- How far in from a fragment's edge are ecosystem processes altered? How can this information be used to estimate the minimum sizes of "intact ecosystems"?

- How do invasions and extirpations of species within fragments alter internal nutrient cycles and carbon storage? Conversely, how does the perturbation of nutrient cycles affect invasions and extirpations?

- What are the spatial and temporal scales of major natural disturbances (e.g., fires, storms, epidemics, droughts)?

- How does fragmentation affect the rate at which these disturbances occur and their impact on populations within the community?

- In addition to a research program on local, intact, and fragmented communities, we also urgently need a global assessment of the rates and causes of anthropogenic ecosystem fragmentation. Remote sensing technology should be exploited to help in this task.

- In the sea, there is need for information on the fragmentation of adult, juvenile, and nursery habitats and its effects on dispersal capabilities and recruitment.

How do different communities and species respond to different spatial and temporal scales of anthropogenic fragmentation? This question should be approached from a landscape perspective, among others. For example, information is needed on how the environment-species matrix in which the fragments are imbedded influences the fate of the species and communities in the habitat fragments. Also, much more information is needed about the roles of topographic barriers and corridors. Specifically,

- How does the degree of isolation of patches influence the rate of colonization and extinction?

- What are the minimum or optimum topological requirements for landscape corridors, and how do these depend on the particular ecosystem and target species?

B. Effects of Fragmentation on Species Interactions and Community Structure

As explained in Chapter 3, fragmentation may alter critical interactions among species, including, but not limited to, predation, parasitism, other diseases, and mutualisms. The following research priorities require immediate attention.

- To what extent does functional redundancy vs. complementarity exist among the species present in a community? Presumably not all species are equally important to the maintenance of ecosystem structure and function.

- Which species can be lost in a local habitat fragment without triggering the loss of a large number of other species?

- To what extent does the impact of losing a component of an ecosystem vary from community to community?

In recent years, the description of edge effects has emerged as a critical research area in conservation biology. Edges benefit some species but create problems for others. Edges undergo succession Edges facilitate the movement of nutrients and the flow of energy. Edge effects may penetrate a few meters or several kilometers. In any case, a much better understanding of edge processes is essential for the management of habitat fragments.

- How do species respond to edges of different physical characteristics?

- How do edges affect the abiotic characteristics of natural communities, such as temperature, relative humidity, light penetration, and wind exposure?

- How do these abiotic changes translate into biotic changes such as tree mortality, windthrow, and leaffall?

- Temperate-zone edges are often areas of enhanced species richness, while tropical edges appear to be regions of depressed species richness. What ecological factors account for this difference?

- To what extent do anthropogenic edges, such as those resulting from logging or farming, resemble treefall gaps, burned areas, and other natural edges with respect to function and species composition?

- What are the consequences of changing ratios of edge to interior habitat on the structure and functioning of communities?

- How do differences in productivity, including predator abundance, between habitats (across edges) influence the magnitude and extent of edge effects?

- How does the matrix surrounding a habitat patch influence the magnitude and extent of the edge effects within the patch? More generally, should matrix effects be distinguished from edge effects?

- Which edge effects are self-catalyzing (creeping), and which are self-limiting?

- To what extent are edge-related phenomena (e.g., elevated rates of nest parasitism in temperate zones, invasion of alien plants, etc.) affected by size and shape of a habitat discontinuity?

Fragmentation will affect the interactions between species, sometimes dramatically. For example, the removal of a large predator will often lead to an increase in abundance of prey species, including smaller predators. This in turn can lead to the local extirpation of the latter's prey species. Fragmentation will also affect mutualistic interactions.

- How do life history parameters in species that are the targets of conservation change following introduction or removal of predators, pathogens, mutualists, etc?

- How does fragmentation alter patterns of movements of mutualistic microorganisms, pollinators, and dispersers?

As always, we need to know how host parasite systems respond to fragmentation, and whether the responses simply reduce the probability of disease outbreaks (by maintaining host populations below threshold), or whether reduced genetic variability in isolated host populations increases their susceptibility to current or introduced diseases? These questions alone beg a better understanding of the role of genetic variability in host-parasite relationships. The demise of the black-footed ferret in Wyoming is an example of a disease (canine distemper) almost eliminating a population of an endangered species that was restricted geographically to a single patch.

- What are the size thresholds of host populations that determine the ability of pathogens or parasites to persist?

- How does fragmentation affect the transmission of different kinds of disease/parasite transmission?

- Does reduced genetic variability in isolated host populations increase their susceptibility to current or introduced diseases?

Another set of problems concerns the genetic structure and gene flow patterns of plant populations.

- How does the size and isolation of patches affect movement patterns of pollinators/dispersal agents and, hence, the genetic structure of plant populations?

- Are specialized mutualists more likely to disappear from fragmented habitats, thereby increasing the dependence of plants on less reliable generalists?

C. Effects of Fragmentation on Single Species

The effects of fragmentation on individual species relative to their persistence requires studies of dispersal mechanisms and abilities, reproduction, and the effects of edges on critical species.

- To what extent are natural and anthropogenic disturbances similar or dissimilar in the species' responses they evoke?

- How can we predict the response(s) of animals to changing

proportions and availability of different habitat types within their home ranges?

- How do the population dynamics of plants and animals change in relation to habitat parameters such as canopy closure and leaf area index that can be measured by means of remote sensing?
- What is the relationship between a species' dispersal abilities and its metapopulation survival?

- How do behavioral mechanisms influence habitat choice and, hence, metapopulation survival?

- How does the composition of the matrix outside a habitat patch influence the vital rates (mortality, fecundity, etc.) of species within the patch?

- How do life history, body size, and other traits affect a species' vulnerability to fragmentation? For example, is the intrinsic rate of growth or its variability more important in determining persistence in an isolated patch?

- How does fragmentation of coastal marine populations affect larval ecology and recruitment?

Fragmentation, through its effects on ecosystem processes, population dynamics, and species interactions, influences the evolutionary pathways of species. Hence, patch populations may acquire or lose characteristics that increase their probability of survival in such a mosaic landscape. Research is needed to better understand how "always rare" and "newly rare" species differ in their life history characteristics and in their potential for persistence in a fragmented landscape.

- Are dispersal patterns and dispersal abilities different among species adapted to extensive vs. fragmented habitats?

- Are species that are adapted to fragmented habitats better at escaping from predators or from interspecific competitors than species that are common in extensive habitats?

- Are species that are adapted to fragmented habitats better able to find mates than species adapted to extensive habitats?

Sri Lanka

Chapter 7
Ethnobiology and
Genetic Resources

Ethnobiology is the study of how different societies interact with the natural environment. Major topics include ways that different cultures perceive, classify, and evaluate biological resources, and ways that societies organize ecosystems for their own needs. The great heterogeneity of human and biological communities is an important resource for both conserving diversity and improving human livelihood. Accordingly, comparative research on how biotic resources are used, maintained, and changed by different societies is useful for developing general theories and methods for managing and conserving these resources.

Although ethnobiology does not conventionally address the conservation of biological resources, there is an emerging convergence between these fields. These linkages are compelled by rapid demographic and environmental changes. Rapid population growth, migration, changing consumption patterns, the diffusion of information and technology, increased cultural contact, and economic development programs have fundamentally altered both knowledge and land-use systems. The populations of many cultures and language groups have shrunk to precariously small sizes; biological resources have become commodities; and land necessary for extensive subsistence strategies has been drastically reduced. Many human groups that once maintained biological resources have either been replaced by other groups or now overexploit their resources.

The implication of these changes is threefold:

1. Ethnobiology must give more emphasis to dynamic and open models of human knowledge systems and behavior that include factors that were considered exogenous in earlier models. An inevitable consequence will be to make ethnobiological research more inclusive, complex, and messy.

2. Ethnobiology must demonstrate its relevance to conservation biology. Such disciplinary integration will require that ethnobiology become more eclectic in its theories and models of human behavior.

3. It is necessary to confront the development-conservation polarity, both in basic and applied research and in policy making. A holistic field such as ethnobiology should be a source for the much needed theoretical synthesis of ecology and human behavior.

One of the most important questions that could be illuminated by ethnobiology is "under what conditions does an individual or a society conserve biological diversity?" It is clear that any useful answer will depend on the cultural context.

While sustainable resource use and conservation of biological diversity per se are widely accepted as mutual goals in technological societies, achieving a balance between them has largely failed. There is a conspicuous lack of examples showing how both traditional and modern societies perceive, value, and conserve biological diversity while successfully using natural resources in a sustainable manner. Ethnobiological research is needed to design more effective and locally acceptable conservation and management plans.

It has long been recognized that all cultures and languages organize the natural world in taxonomic ways. The universality of this human ordering of the natural world presents a number of important questions concerning the nature of knowledge systems. These include:

• In what ways do specific cultures or societies perceive and classify biological resources?

• How does this classification affect the evaluation and use of biological resources?

- How do perceptions and evaluation criteria change over time and in different contexts?

A major contribution of inquiries of this sort has been to document the complexity of indigenous knowledge systems. Non-Westion systems are often as ordered and detailed as are Western scientific ones. A second contribution has been to describe how some human production systems in areas of great biological diversity do not adversely affect diversity. Renewable resources have been conserved by maintaining low population densities and limiting consumption. Some traditional practices, can be used as models. For example, traditional Polynesian fisheries conservation practices have been adopted in contemporary management schemes.

Priority Areas of Research

A. The Effects of Social and Economic Change on Biological Diversity

A common framework of biological and social science research is needed. Three categories of change should be addressed: (1) increasing population pressure through growth and migration, (2) changes in expectations and consumption, and (3) the integration of a particular locality or region into market systems of different scales. Research questions include:

- How do groups extrapolate the future use of biological resources? Mathematical models predicting rates of use must depend on both time-series and cross-sectional databases. In addition the rate of habitat loss and species extinction must be better understood.

- How do use rates change as biological, social, economic, and cultural factors change?

- What information do farmers, conservationists, corporations, politicians, and other groups need about the biological resources which they affect?

B. Land-Use Patterns and Biological Diversity

Conservation depends on a mosaic of land-use systems. It is essential to develop information on and methods of assessing the impact of agricultural intensification in certain production zones. There is always the danger that intensification can exacerbate habitat fragmentation (see Chapter 6). Environmental and social components determine the mosaic of land and water use. Therefore, we need to know:

- How do different land management systems, such as individual property and common property, affect the use and protection of natural resources?

- How does the size of land management units affect the loss/protection of biological diversity?

- What combinations of extensive and intensive land use achieve satisfactory sustainable returns to people while conserving biological diversity?

C. Systems Level Research and Land-Use Patterns

Although it is difficult to identify researchers doing research on "unsustainable" systems, a widely held notion is that discipline-specific (or reductionist) research has led to technologies and policies that are detrimental to biological resource bases. Granting that the whole of biological and social science research may be construed as leading to a general systems understanding, there is need for more explicitly systems level research. Research questions in this area include:

- What anticipated returns (economic, energetic, informational, aesthetic) are used in different land management systems, and how can these be combined to alter management?

- Can single system properties, such as net energy balance or information storage, be used by both biological and social science to improve land management for conservation purposes?

D. Designing New Production Technologies and Land-Use Systems

Increased pressure on biological resources arises because of increasing human populations, changing consumption patterns, and new technologies. Although agricultural intensification will continue to be necessary, its impact on biological resources is not predetermined. Conservation poses important research questions relevant to the design of new production technologies and land-use systems.

- Can biologically diverse and low energy technologies (e.g., paddy cultures, chinampas, house gardens) be extended and/or intensified?

- Can production systems be differentially intensified so as to maintain biological diversity in other parts of a system?

- How does increased exploitation of specific species affect other species and general system properties? (See Chapter 3).

E. Relating Economic Return to Conservation Values

Parks and reserves are essential components of any conservation strategy, but it is clear that such reserves inevitably will be too small to maintain species, ecological interactions, and evolutionary processes at satisfactory levels. Therefore, some scientists have argued that an important part of the conservation "burden" must be borne by those lands and waters that are being exploited for economically valuable products. Traditionally these areas have been managed almost exclusively to maximize the economic returns from them, with little concern for the effects of that management on species other than those of direct economic concern. Mixed-use strategies, such as agroforestry, require research that relates economic return to conservation value. Several important questions are:

- Do mixed species plantings increase species richness significantly? To what extent does this depend on the geographic region, latitude, etc.? Which species groups are enriched?

- How does management, such as harvest schedules and no-tillage agriculture, affect plant and animal species richness?

- Is it economically viable to manage forests for both mixed species and mixed-age composition rather than for single species, single-age stands that are currently favored? If so, what are the consequences for species richness?

- What economic conditions favor the survival of species by creating private property rights for species and their economically valuable products?

- How can species richness be maintained on exploited lands that are adjacent to parks and reserves?

- How do traditional pastoral systems help to maintain plant and animal species?

F. On-Site Conservation of Germ Plasm Resources

Food and fiber production in managed ecosystems depends on periodic infusions of genetic materials. Much of the genetic diversity that supports agriculture worldwide originates in the tropics and in landscapes managed by small-scale farmers practicing traditional agriculture. Social and economic changes threaten genetic resources in the same ways that they threaten wild ecosystems.

The conservation of genetic resources of economically useful plants and animals can be accomplished best by a combination of on-site and off-site strategies (botanic gardens, zoos, etc.). Although off-site conservation may be appropriate and necessary for some species, on-site conservation is essential for others. For many tree species, on-site conservation is necessary. For crop species, the conservation of wild and weedy relatives and ancestral crop varieties ("land races") now depends on off-site conservation. On-site conservation of crop resources may be more effective from both economic and biological perspectives, but it is politically difficult to establish. It may require the active participation of local populations, and it will depend on both institutional and popular support.

Ethnobiological research can enhance off-site conservation by increasing our understanding of patterns by which some genetic resources are lost and others are maintained so that collection may be accomplished more efficiently. Ethnobiology can also help in the design of on-site conservation projects by identifying why and how certain regions or human groups conserve biological diversity in their agriculture and other land-use practices. Research questions include:

- What are the biological and economic costs and benefits of on-site and off-site conservation?

- Where can on-site methods be used as a type of complementary strategy to off-site methods?

- What ecogeographic and socioeconomic information on patterns of environmental change is necessary to plan on-site conservation?

- What social, economic, and political factors at both the local and national levels are supportive of and contrary to on-site methods?

71

Calypso bulbosa

Chapter 8
Diversity:
Inventory and Systematics

A knowledge of diversity is a basic prerequisite of conservation biology. Without quantitative measurements of diversity, based necessarily on inventory and systematics, it is not possible to produce rational conservation policies, especially in the species-rich areas of the tropics. Measurements of diversity are dependent upon the basic knowledge of species and their distributions in space. These fundamental data are the foundation of all the biological sciences. Biogeography, the study of past and present distributions of species, is essential in management because it addresses such problems as commonness and rarity, habitat requirements, and barriers to dispersal.

The inventory of the world's biota is far from complete. Biologists can say roughly how many birds, mammals, and flowering plants there are on each of the continents, but a continental scale is too broad to be of any use in conservation planning. Existing parks and reserves mostly lie in the range of 100 to 10,000 square kilometers; at such a scale biological information is rarely available. Existing inventories usually cover much smaller areas, such as the flora of Barro Colorado Island, Panama, covering 16 square kilometers, or the flora of the Palenque Biological Station in Ecuador, covering only one square kilometer. Yet these are too small to take into account the diversity of vegetation communities inevitably contained in a watershed or other landscape unit appropriate for preservation as a reserve.

There is thus a pressing need for surveys and mapping to resolve the spatial pattern and scaling of biological diversity in each of the world's principal biogeographic regions. For the tropics, particularly the humid tropics, where most biodiversity is found, a crash program is needed to identify and map the major spatial patterns of diversity on appropriate scales. Only through such a concerted mapping effort will biologists be able to locate centers of endemism, diversity "hot spots," and distinctive plant formations associated with unusual soil or hydrological conditions. Such information is a vitally needed prerequisite for rational planning and for the preservation of biological diversity. Without such information, conservation can only shoot in the dark (see Appendix B).

The description and maintenance of biodiversity is a multidisciplinary challenge. Team research involving scientists from many different disciplines, and representing a wide range of taxonomic specialties, should be encouraged and given funding priority.

Finally, a word of caution is appropriate here. While we must continue the planetary inventory to identify all organisms and to refine conservation policy, conservation cannot wait for its completion. Likewise, estimation of actual species diversity is essential, but debate about which type of measurement is best can also become a distraction. Even our incomplete inventory contains a huge amount of data. Taxonomic coverage is very uneven, but some organisms, such as birds and butterflies, are relatively well known. In order to formulate conservation policy while there is still time to conserve ecosystems, we will have to use the existing data. By analysis of the available information about certain selected organisms, we can certainly formulate a conservation policy that could preserve a large number of species. It is preferable to make policy decisions now on the basis of existing data than to wait while the species under study become known only as museum specimens. We therefore recommend a strategy that would make the maximum possible use of existing data.

Priority Initiatives

A. Identify Critical Areas in Need of Immediate Protection

First and foremost, there is a need to bring together the scientists who hold most of the data on the tropical organisms to pool their information and select the most logical areas for conservation. This can only be accomplished by creating an interactive environment over an extended time period. Either a series of meetings or a month-long meeting should be arranged in 1989 to accomplish this task. The crisis is so great that we cannot wait any longer to voice a unified rationale for conservation.

B. Produce a Set of Handbooks on Tropical Ecosystems, Conservation, and Conservation Biology

In addition, we propose the production of a set of multilingual handbooks summarizing knowledge of natural history, biogeography, and ecosystem processes of tropical regions that is of immediate relevance to conservation and conservation biology. They would (1) provided a framework for current practical conservation efforts (acquisition, species reintroductions, etc.) in a format that would convince policy makers of the data upon which conservation is based, (2) serve as a starting point for new researchers in the tropics, (3) serve as a basic tool for teaching conservation and be useful to specialists who wish to broaden their knowledge, and (4) be a guide to the literature.

Priority Areas of Research

The initiatives outlined above address the immediate need for urgent action, but they will not suffice in the long-term. Further research is needed to refine and extend the site-selection process for conservation action and to enable management of already protected areas. Once areas have been selected, it is important to know the species they contain, their evolutionary relationships, and their distribution globally, regionally, and locally.

It should be obvious that effective action will depend on adequate support for museums and similar institutions. Such support for local and regional institutions in developing countries is essential for the long-term success and continuity of conservation in the tropics. As discussed in the next chapter, each aspect of research should involve the training of local people at all levels.

The inventory and descriptive phase of diversity is far from complete. There is an urgent need to accelerate collection along with descriptive taxonomy by systematists who are willing to consider the role of the organisms which they study in the ecosystem and are able to synthesize their data in such a way that they can be used for conservation. For example, very little is known about such ecologically important groups as free-living nematodes and orobatid mites. Given their ecological significance, much more information is also needed on the systematics and biogeography of social insects.

Current efforts to catalog diversity are too often hindered by hit- or-miss methodology. There is a need for much closer coordination of effort and much greater emphasis on the associated use of modern methods such as computer data banks. Every specimen collected today should be entered into a data bank as its label is typed. Diversity research must include long-term studies. The monitoring of communities over time is an essential component that is severely lacking in tropical research.

- Data cataloging formats should be universal and should be developed by teams of systematists, ecologists, and biogeographers.

- Additionally, there is a need for the development of a standardized methodology (protocols, surveying and sampling techniques) and sets of rules for mapping of vegetation formations and coastal marine zones in tropical regions (the mapping of climates, soils, etc., should also be carried out where appropriate).

- Research should determine the spatial scales at which diversity changes. This includes research on the minimum critical sizes of ecosystems to be conserved.

Multilingual copies of maps and accompanying materials should be available within each country. Such maps will be used to: (1) identify major hot spots of diversity and endemism, (2) identify gradients of diversity, resources, productivity, and disturbance, (3) provide a framework for continued monitoring of changes in tropical forest biomass, diversity, and fragmentation, and (4) make specific recommendations for preservation.

Farm in Beni region, Bolivia

Chapter 9
Infrastructure and Training in the Developing World

Effective conservation cannot be sustained without the presence of three kinds of infrastructure: social, managerial, and scientific. It is widely acknowledged that the ultimate effectiveness of programs for conservation and sustainable development depends on the success of domestic environmental pressure groups, including grass roots organizations. Private voluntary organizations provide the political momentum and continuity that international organizations and lending institutions can initiate but cannot sustain indefinitely.

The justification for an emphasis on educational programs in the developing world includes the following points:

1. Most of the world's species occur in tropical areas, and the tropics and the developing world are virtually synonymous.

2. The rate of species and habitat loss is much higher in tropical than in temperate areas. In the tropics, conversion of natural habitats to agricultural land is differentially affecting certain habitats and their biological communities. For example, coral reefs are disproportionately represented in tropical countries, and reefs are especially vulnerable to human activities.

3. The citizens of these tropical countries will be primarily responsible for the long-term conservation of biological diversity. Their success

in this critical enterprise will depend on political will and an institutional base (teaching, research, and administration) for the protection and conservation of biodiversity and natural resources in their own countries.

4. Such a constituency and such an infrastructure cannot evolve in the absence of education in the principles of conservation and sustainable development.

5. This, in turn, will depend on training in approaches to conservation biology and on the development of policy that encourages conservation.

Priority Initiatives

A. Fortifying Private Voluntary Organizations

Funders should encourage the development of short courses, workshops (certificate programs), and graduate programs for training nationals of developing countries in the formation and operation of private voluntary organizations within their own cultural contexts. The curricula of such programs would contain the scientific fundamentals of conservation; analytical skills including various approaches to decision making and the use of microcomputers; organizational development skills including resource development, staff recruitment and training, and strategic planning; influence skills such as lobbying, negotiation, conflict management, and the politics and sociology of government agencies; and interorganizational skills. Sensitivity of the students to the unique opportunities and qualities of their country's institutions and cultures should be cultivated. Funders, including government agencies, should also provide fellowships and scholarships to support students from developing countries.

B. Short Courses and Graduate Education in the Relevant Scientific Fields

Three successful approaches to graduate education can be identified: (1) short courses (1-3 months) in both temperate and tropical settings, (2) graduate training at universities in the developing world, and (3) graduate training at universities in the developed world.

Short Courses

Short courses are an effective means of providing training to a large number of students. Large numbers of nationals from developing countries need to be trained in the approaches of conservation biology if tropical ecosystems are to be conserved and managed. Two types of short courses can be easily distinguished: those aimed at senior policy makers and senior professionals and those aimed at resource managers, administrators, wardens, and park personnel (see Section C below). Both provide an opportunity for future networking, and both help people to do their jobs better. Such courses also offer a great deal of information and techniques to trainees. What these courses do not do is encourage trainees to think about conservation problems in innovative ways; nor do they provide long-term support for trainees once they return to their former positions. The following kinds of programs are more likely to provide these skills and opportunities.

Graduate Training At Universities in the Developing World

Graduate training at universities in developing countries is gaining acceptance. Unlike the short courses, these graduate programs provide students with a theoretical framework that allows a broader vision of conservation issues and the development of innovative solutions to problems, both theoretical and applied. Through the relationships established during his or her studies, a student might receive considerable support, both conceptual and logistical, once training is completed. These programs tend to have a national or regional focus,

81

which is a strength in that language and cultural problems are minimized, but a potential weakness in that the education might be more
parochial.

Graduate Training at Universities in the Developed World

Graduate training at universities in developed countries generally
provides the most rigorous education in disciplines relevant to conservation biology. Academic institutions defining conservation problems
and developing innovative answers to these problems are mostly found
in the developed world. At these institutions, students receive a broad,
frequently multidisciplinary, education in conservation biology. Students are exposed to a broad range of conservation issues, both pure and
applied. Once they return to their own countries, graduates of these
programs can exploit the personal and academic ties they established
with colleagues, with government agencies, and with non-governmental organizations. It is the graduates of these programs who are the most
likely to set up training programs, graduate curricula, and research
programs in conservation biology in their own countries. A common
weakness of such programs is that they lack relevance to the culture and
environment of students from developing world countries. Innovation
is called for here.

C. Training Programs for Managers: Funding Needs and Criteria

Different developing world countries have different needs. For
some, training a cadre of people who can act as managers and administrators for natural areas is essential. For others, training people who
will in turn establish the educational and training centers in their own
countries is more important.

In order to be effective, managers must be part of the social fabric,
in part to insure their long-term survival. Hence, not only is there a
critical demand for regional institutes or schools for the training of
lower- to mid-level managers of conserved wildlands, but also the
subject matter of such "schools" needs to speak to attitudes (e.g., why

am I, a local farmer employed to be a site manager in a national park, doing this?) as well as to practical skills such as driving a car, constructing an annual budget, and balancing the social benefits of allowing controlled hunting in a national park against the biodiversity losses. The *modus operandi* of the "school" also need to range from actual classroom courses to intra-tropical internships and apprenticeships. A major stumbling block in the growth and development of such programs is the tendency for developing countries to want to import the "student" (and to treat him/her as a "student") rather than being willing to bring the school to the participant and recognize the other social demands on him/her as a mature adult with a key role in society. Funding organizations can meet these needs in two ways:

- They can fund the development and improvement of programs that provide the appropriate training. This "institution-building" approach is most realistic for the short courses for training lower- and mid-level managers and administrators.

- The second recommendation is to provide funds to channel existing and future managers into these courses and to allow individuals to participate in these opportunities.

The latter can be accomplished by providing funds directly to individuals through fellowship programs, or indirectly by encouraging investigators to include support for tropical country collaborators in their research budgets. This second approach is very appropriate because most research projects in conservation biology ultimately require application to real world situations, and this application in the tropics requires the committed participation of informed citizens from tropical countries.

D. How Funders of Research Projects Can Facilitate Training

Funders have a unique opportunity to foster graduate education and training by specifying criteria for research proposals. Although such criteria need not become absolute requirements, they would be used in the evaluation process. We suggest the following:

- Research in conservation biology in tropical countries should involve the active participation of institutions in these countries. By encouraging active involvement, funders will foster interest and participation at policy and managerial levels.

- Investigators from industrialized developed countries should be encouraged to include support of students from tropical countries in technical roles. In this way, investigators will train students in biological techniques and encourage an awareness and interest in biological systems. This apprenticeship may lead the student to seek further education in a discipline relevant to conservation biology.

- Investigators also should be encouraged to include funds for fellowships in research proposals for projects that would support graduate training of students from tropical countries. Promising graduate candidates could be identified by individual investigators and through participating institutions in the country.

- National Science Foundation (NSF) and other funders should provide funds to directly support a fellowship program that would support graduate studies in the United States for students from tropical countries. Criteria used in evaluating candidates should include the applicability of the candidates' proposed research to real situations and the feasibility of such an application. We recommend that investigators applying for fellowships be actively involved in conservation research, based at U.S. institutions.

- NSF and other funders should earmark funds to directly support research in conservation biology by graduate students from tropical countries in their own countries.

- Funders should earmark funds to support research by recently graduated students from tropical countries in the first few years following their return to their own countries. This would help these young researchers through a critical time in the establishment of research/training programs.

Appendix A
Land Acquisition and NSF Funding

There are many ways to acquire land, whether for the purpose of immediate conservation or for research in conservation biology. Land can be purchased outright or donated. Public lands can also be assigned to conservation categories by policy decisions and zoning procedures. Which kind of land acquisition is appropriate for a given region or specific conservation or research project depends on both the biological properties of the site and the social circumstances in the region. There will be times when simply buying a piece of land to protect it will produce significantly more long-term, positive research (as well as conservation) results than would spending the same amount of money on research per se at a given moment.

The need for experimental sites cannot be met simply by manipulating land already conserved in natural areas. Managers of conserved wildland are usually not interested in letting the nature they have fought and bled for be manipulated in the fairly violent manner necessary for many kinds of conservation experiments. While much low-impact, short-term conservation biology research can be carried out on lands that are already firmly secured, certain kinds of research demand the acquisition of large areas. These areas not only allow the research to occur, but are conserved as a by-product.

Land acquired for large-scale or long-term research in conservation biology is generally available for a great variety of other kinds of (often esoteric) field research—research that is often synergistic with the conservation research. Conversely, a national park or other conserved wildland is in fact a biological research station (but has a much more complex administrative demand owing to its larger size and more diverse user community). Land acquisition is preferable to research in some circumstances.

The challenge is different in the marine realm. While the sea cannot be purchased, it can be protected, albeit with greater difficulty than the land. An important conceptual approach is "regional seas," or large marine ecosystems. Funding is needed to establish regional networks of marine laboratories that can serve both as centers for the study of phenomena at the appropriate geographic and temporal scales and as environmental watchdogs.

Megaexperiments

Conservation biology, and particularly tropical conservation biology, is in desperate need of "megascale" (requiring large areas or being of long duration) field experiments, virtually all of which require four conditions: (1) wildland acquisition, (2) time, (3) long-term budget commitment, and (4) local and national integration of the project.

Specific examples of such conservation experiments include:

1. Minimal critical area trials, including the impact of equal habitat reduction on different kinds of habitat;

2. Landscape restoration that has various objectives and uses various methodologies;

3. Configuration trials (i.e., acquisition of experimental corridors, mosaics of land-use types);

4. Wildland captive breeding, reintroduction, augmentation of "natural" species richness, etc.;

5. Selective species harvest or enhancement (through experimental removal, supplemental feeding);

6. Comparisons of boundary designation vs. "natural boundaries" (e.g., on ridges, rivers, lakeshores, etc.);

7. Analyses of user impacts on conserved areas;

8. Comparing management techniques (to burn or not to burn, use of cattle as biotic mowing machines, introduction of biological control agents for foreign weedy species, etc.).

Among the other three requisites, the time element is probably the least problematic since interim experimental results will become apparent almost immediately, and these results will aid in the interpretation of nearby "natural" experiments (e.g., pastures abandoned for varied numbers of years). The social integration of the project is essential and is a research area with great potential in the social sciences. Topics of obvious importance include conflict resolution, occupancy and relocation, and community politics. Research that relates the category of land acquisition to social issues is also needed. We also need cost/benefit analyses and risk assessment studies of the different forms of land acquisition that can be applied to a given site.

Of the four requisites mentioned above, it is the start-up capital for land acquisition and physical plant development that is often severely limiting. Land acquisition is a very specific need. It is appropriate and economically fair that researchers contribute to the capital start-up costs through land purchase and to wildland maintenance costs. The National Science Foundation should view land purchase and maintenance in exactly the same way that it views the purchase of a piece of fancy machinery, setting up a lab or animal care center, insurance, preventative maintenance, repairs, and power (diesel, electric bills, etc.). If there are legal barriers to direct acquisition of land in other countries by U.S. government agencies, then alternatives such as grants to such countries for the establishment and management of research reserves should be explored. A potential funding source would be Public Law 480 programs which are currently operating in many developing countries.

Grant proposals to a NSF Program in Conservation Biology that include land acquisition in one form or another require analytical abilities of a type not well represented in the academic community. This suggests a different mode of proposal review, one that involves representation from the host country and/or the administrative unit that will be most directly involved with the land acquired. In addition, representation would be desirable from the conservation nongovernmental organization (NGO) community that already has considerable experience in land acquisition for conservation purposes.

Guidelines would have to be developed insuring that ownership of the land be vested in the host country's conservation system (national park service, national forest service, etc.). There exists precedent for this.

Land acquisition for conservation research is not nearly as expensive as it might appear. $100,000 purchases a piece of scientific hardware. $100,000 may purchase a thousand to a million hectares of tropical wildlands, if that initial funding is manipulated appropriately. As for value received, the hardware ends up in the dumpster in less than ten years; the wildland can persist forever and will be used by many generations of researchers.

Appendix B
The Smithsonian BIOLAT
Program for Biological Survey

An increase in the rate of biotic inventorying will put tremendous new burden on taxonomists. New and flexible taxonomic techniques will be called for. In addition, it will be necessary to develop an interim taxonomy for poorly known yet important groups of organisms, such as most arthropods. New procedures will be required to associate the results of surveys and inventories to vouchered but undescribed specimens and species. Creative use of computers and other techniques will make this possible and will avoid the necessity of multiplying the number of taxonomists by one or two orders of magnitude. Nevertheless, there should be a modest increase in the number of taxonomists using such approaches, particularly in tropical countries where most biological diversity exists.

The Smithsonian Institution has developed protocols, the "BIO-LAT Program," that embody these concepts for inventorying little known tropical areas and taxa. This integrated inventorying approach is based on recording the co-occurrence of plants and animals in permanent plots by utilizing field computers and software tailored to producing an interim taxonomy. The near-term objectives of the program are: (1) to establish permanent plots in all representative plant communities in a network of protected areas along the western and northern Amazon Basin, and then to expand this system throughout the Neotropics, (2) to inventory all species of plants, animals, and microorganisms in these plot,; (3) to provide checklists of species in each protected area, and (4) to monitor changes, in species population numbers in each permanent plot in relation to plant community dynamics. The protocol is being produced in a communicative format in order to enhance its utility in training, research, and in resource management.

Information on the BIOLAT Program can be obtained through Terry L. Erwin, BIOLAT-SI/MAB Biological Diversity Program, National Museum of Natural History, Smithsonian Institution, Washington, D.C. 20560.

Appendix C
Workshop Participants

Carol Augspurger
University of Illinois
Department of Plant Biology

James Brown
University of New Mexico
Department of Biology

Wesley Brown
University of Michigan
Biology Department

Stephen Brush
University of California, Davis
International Agricultural
Development

Martin Caldwell
National Science Foundation

Jeremy Cherfas
New Scientist Magazine

Harold C. Conklin
Yale University,
Department of Anthropology

Andrew Dobson
University of Rochester
Department of Biology

Paul Ehrlich
Stanford University
Department of Biological
Sciences

Winifred Hallwachs
Univesity of Pennsylvania
Department of Biology

W. Franklin Harris
National Science Foundation
Biological, Behavioral,
and Social Services

Philip Hedrick
Pennsylvania State University
Department of Biology

Steven Hubbell
Princeton University
Biology Department

Michael Huston
Oak Ridge National Laboratory

Daniel Janzen
University of Pennsylvania
Biology Department

Kathleen Keeler
University of Nebraska
School of Biological Sciences

Timothy Lawlor
National Science Foundation
Systematic Biology Program

Thomas Lovejoy
Smithsonian Institution

John Ogden
Florida Institute of
Oceanography

Gordan Orians
University of Washington
Zoology Department

William Perrin
NOAA
National Marine Fisheries

Ghillean Prance
Kew Botanical Gardens

Katherine Ralls
Smithsonian Institution
National Zoo

John Robinson
University of Florida
Forest Resources and
Conservation

Ulysses Seal
U.S. Veterans Administration
Medical Center, Minneapolis

Mark Shaffer
U.S. Fish and Wildlife Service
International Affairs Office

Daniel Simberloff
Florida State University
Biological Sciences

Melvin Simon
California Institute of Technology
Biology Division

Michael Soulé
University of Michigan
School of Natural Resources

Laura Tangley
BioScience Magazine
American Institute of
Biological Science

John Terborgh
Princeton University
Department of Biology

James Tucker
Goddard Space Flight Center

Christopher Uhl
Pennsylvania State University
Department of Biology

David Western
Wildlife Conservation
International, New York
Zoological Society

Walter Westman
University of California
Lawrence Berkeley Laboratory

David Wilcove
The Wilderness Society

David Wildt
Smithsonian Institution
National Zoo

Further Readings in
Conservation Biology

Diamond, J.M. 1975. The island dilemma: lessons of modern biogeographic studies for the design of natural reserves. *Biological Conservation* 7: 129-46.

Ehrlich, P.R. and A.H. Ehrlich. 1981. *Extinction: the causes and consequences of the disappearance of species.* Random House, New York.

Harris, L.D. 1984. *The fragmented forest: island biogeography theory and the preservation of biotic diversity.* The University of Chicago Press, Chicago, Illinois.

Harris, L.D. 1988. The nature of cumulative impacts on biotic diversity of wetland vertebrates. *Environmental Management* 12: 675-693.

Myers, N. 1988. Threatened biotas: "hotspots" in tropical forests. *The Environmentalist*: 1-200.

NAS (National Academy of Science). 1980. *Research priorities in tropical biology.* National Academy of Science, Washington, D.C.

Newmark, W.D. 1987. A land-bridge island perspective on mammalian extinctions in western North American parks. *Nature* 325: 430-432.

Office of Technology Assessment. 1987. *Technologies to maintain biological diversity.* OTA-F-330. U.S. Government Printing Office, Washington, D.C.

Pickett, S.T. A. and P.S. White (eds.). 1985. *The Ecology of Natural Disturbance and Patch Dynamics*. Academic Press, Orlando, Florida.

Schonewald-Cox, C.M., S.M. Chambers, F. MacBryde, and L. Thomas (eds.) 1983. *Genetics and conservation: a reference for managing wild animal populations.* Benjamin/Cummings, Menlo Park, California.

Soulé, M.E. and B.A. Wilcox (eds.). 1980. *Conservation biology: an evolutionary-ecological perspective.* Sinauer Associates, Sunderland, Massachusetts.

Soulé, M.E. (ed.) 1986. *Conservation biology: the science of scarcity and diversity.* Sinauer Associates, Sunderland, Massachusetts.

Soulé, M.E. (ed.) 1987. *Viable populations for conservation.* Cambridge University Press, Cambridge, Massachusetts.

Terborgh, J. 1975. Faunal equilibria and the design of wildlife preserves. Pp. 369-380 in F. Golley and E. Medina (eds.). *Tropical ecological systems: trends in terrestrial and aquatic research.* Springer-Verlag, New York.

Wilson, E.O. (ed.) 1988. *Biodiversity.* National Academy Press, Washington, D.C.